T0191964

Analog Circuits and Signal Processing

Series Editors
Mohammed Ismail, Khalifa University, Dublin, OH, USA
Mohamad Sawan, Montreal, QC, Canada

The *Analog Circuits and Signal Processing* book series, formerly known as the *Kluwer International Series in Engineering and Computer Science*, is a high level academic and professional series publishing research on the design and applications of analog integrated circuits and signal processing circuits and systems. Typically per year we publish between 5-15 research monographs, professional books, handbooks, and edited volumes with worldwide distribution to engineers, researchers, educators, and libraries.

The book series promotes and expedites the dissemination of new research results and tutorial views in the analog field. There is an exciting and large volume of research activity in the field worldwide. Researchers are striving to bridge the gap between classical analog work and recent advances in very large scale integration (VLSI) technologies with improved analog capabilities. Analog VLSI has been recognized as a major technology for future information processing. Analog work is showing signs of dramatic changes with emphasis on interdisciplinary research efforts combining device/circuit/technology issues. Consequently, new design concepts, strategies and design tools are being unveiled.

Topics of interest include:

- Analog Interface Circuits and Systems;
- Data converters;
- Active-RC, switched-capacitor and continuous-time integrated filters;
- Mixed analog/digital VLSI;
- Simulation and modeling, mixed-mode simulation;
- Analog nonlinear and computational circuits and signal processing;
- Analog Artificial Neural Networks/Artificial Intelligence;
- Current-mode Signal Processing;
- Computer-Aided Design (CAD) tools;
- Analog Design in emerging technologies (Scalable CMOS, BiCMOS, GaAs, heterojunction and floating gate technologies, etc.);
- Analog Design for Test;Integrated sensors and actuators;
- Analog Design Automation/Knowledge-based Systems;
- Analog VLSI cell libraries;
- Analog product development;
- RF Front ends, Wireless communications and Microwave Circuits;
- Analog behavioral modeling, Analog HDL.

More information about this series at http://www.springer.com/series/7381

Chung-Chih Hung • Shih-Hsing Wang

Ultra-Low-Voltage Frequency Synthesizer and Successive-Approximation Analog-to-Digital Converter for Biomedical Applications

 Springer

Chung-Chih Hung
Department of Electrical
and Computer Engineering
National Yang Ming Chiao Tung
University
HsinChu, Taiwan

Shih-Hsing Wang
Memory Design and Analysis Division
Etron Technology, Inc.
HsinChu, Taiwan

ISSN 1872-082X   ISSN 2197-1854   (electronic)
Analog Circuits and Signal Processing
ISBN 978-3-030-88847-3   ISBN 978-3-030-88845-9   (eBook)
https://doi.org/10.1007/978-3-030-88845-9

This Springer imprint is published by the registered company Springer Nature Switzerland AG
The registered company address is: Gewerbestrasse 11, 6330 Cham, Switzerland

Preface

For humans to complete all actions in life, we must transmit biomedical signals through corresponding nerve fibers to the central nerves of the brain and tissues to achieve all activities and functions. Collecting the corresponding biomedical signals can not only continuously grasp the biomedical conditions of the human body, but also issue timely warning signs of the symptoms before onset.

Recently, mobile phones have combined with various wearable devices to monitor or measure the relevant physiological information of the human body. However, engineers are not familiar with many physiological signals. Therefore, in the first chapter, we briefly introduce the origin of some biomedical signals and the working principles, measurement, and subsequent processing behind them. Chapter 2 discusses how to design low-power circuits. Since these wearable devices, sensor devices, or implanted devices are powered by batteries, they need to maintain a long working time. We hope to reduce their power consumption to extend their service life, especially for implanted devices because the battery needs to be replaced by surgery. Both implantable and in vitro medical signal detectors require two basic components to collect and transmit these signals: an analog-to-digital converter and a frequency synthesizer because these measured biomedical signals are wirelessly transmitted to the relevant receiving unit. The most important element in the wireless transmission is the frequency synthesizer. The frequency synthesizer that provides a stable frequency is the core unit of the wireless transmitter. Therefore, Chap. 3 mainly introduces the basic working principles and models of frequency synthesizers. In Chap. 4, we provide a design example and measurement results of a low-power, low-voltage integer-N frequency synthesizer for biomedical applications. The detection of biomedical signals needs to be converted into digital signals by an analog-to-digital converter before relevant signal processing and subsequent judgments can be carried out. Therefore, Chap. 5 introduces the working principles of the analog-to-digital converter. It also briefly introduces many types of analog-to-digital converters. In Chap. 6, we show an implementation example and measurement results of a low-power, low-voltage analog-to-digital converter for biomedical applications. Chapter 7 summarizes the book.

In the near future, more and more biomedical signals will be seamlessly captured through electronic hardware and software. It can help normal people control their weight and maintain good physical condition. It can also help people with physical illnesses to enhance their quality of life. We believe understanding biomedical signals can bring benefits to people. Hopefully, this book can provide guidance to readers who are interested in the design of biomedical electronic devices.

HsinChu, Taiwan   Chung-Chih Hung
HsinChu, Taiwan   Shih-Hsing Wang
August 2021

Contents

Chapter 1
Introduction to Biomedical Signals and Their Applications

For majors in electronic or electrical engineering, the origin and characteristics of biomedical signals are not particularly understood. Thus, in this chapter, we briefly introduce the origin and transmission of some biological signals, as well as common biological signal characteristics, including amplitude and frequency. The biological signal may introduce some noise during the measurement. Therefore, noise must be eliminated first when processing biological signals. Feature recognition can be obtained through signal processing procedures, such as dimensionality reduction or signal domain conversion. Some information brought by biological signals can be used for special purposes. For example, rehabilitation aids or equipment can be fabricated through the human–machine interface to improve the quality of life. The implanted device can also be used for patients to immediately apply appropriate feedback, such as cochlear implants to stimulate the auditory nerve or the pacemaker to stimulate the heart to beat; these signals can also be wirelessly transmitted to the medical center or emergency center to notify the relevant units to respond and immediately prepare so that the patient can receive treatment and immediate care.

1.1 Introduction to Biomedical Signals

Biomedical signals are mainly used to diagnose or detect specific pathological or physiological conditions. These signals are also used to analyze biological systems in healthcare, research laboratories, clinics, and even homes. Common examples of biomedical signals are electrocardiogram, electroencephalogram, and electromyography. These biomedical signals are used to identify information about complex pathophysiological mechanisms and to diagnose different diseases. However, the original biomedical signal has additive noise and cannot be directly analyzed. It requires biomedical signal processing to enhance relevant information and remove noise, including denoising, filtering, spectrum estimation, and feature extraction. An accurate signal model can then be created to analyze its components and predict pathological events in the brain, heart, or muscle.

1.2 Origin of Biosignal or Bioelectrical Signal Biopotential

A biological signal is any signal that can be continuously measured and monitored in the body. The term biosignal is generally used to refer to bioelectrical signals/ potentials, but it can also refer to electrical and non-electrical signals. Usually, it only refers to the signal that changes with time and sometimes also includes the change of spatial parameters.

Generally, if we consider the human body as a source, then the source of the electrical potential generated from/through the cell area will only be due to the concentration of ions passing through the cell membrane. All cell membranes have an envelope (selectively permeable or semipermeable) that can regulate ions (such as sodium (Na^+), potassium (K^+), calcium (Ca^{2+}), magnesium (Mg^{2+}), and chlorine (Cl^-)) to maintain the cell potential.

When excited, the nerves and/or muscles contract; this is due to the development of action potentials across the cell membrane. The development of action potentials is the basis or main requirement for the development of bioelectrical signals to utilize the physiological information of specific areas (nerves, muscles, organs, etc.). What is important is that the development of the action potential is due to changes in ion concentration in that particular area or part. This is due to the development of the potential difference.

In short, the cell membrane is surrounded by cytoplasm, which contains "n" proteins and nucleic acids (i.e., biomolecules). The membrane also acts as a protective layer by isolating/separating the cells from the surrounding environment.

The cell membrane is usually composed of a double layer of phospholipids (including lipophilic, hydrophilic, and amphiphilic). Therefore, the phospholipid bilayer in the membrane helps/prevents the movement of ions/molecules in and out of the cell. Due to changes in the concentration of ions inside and outside the cell (such as sodium (Na^+), potassium (K^+), and chlorine (Cl^-)), the muscles and nerves will be at the cellular level (i.e., bioelectrical signals/potentials), as presented in Fig. 1.1. Potential differences can also be produced/developed through electrochemical changes, which are related to the conduction of neurons in and out of the brain in the form of neurotransmitters and receptors. The generated signal and artifact/noise amplitude are very small (typically −90 to +20 mV) [1].

The cells are separated from the surrounding ionic body fluid by a semipermeable membrane. Only through these ionic body fluids can bioelectrical signals (especially neuron/synaptic information) be transmitted between one cell and another cell or a group of cells to perform specific functions. While stimulating a specific cell, the cell membrane of a specific cell will voluntarily let potassium (K^+) and chlorine (Cl^-) ions flow into the cell while restricting sodium (Na^+) ions from passing through the

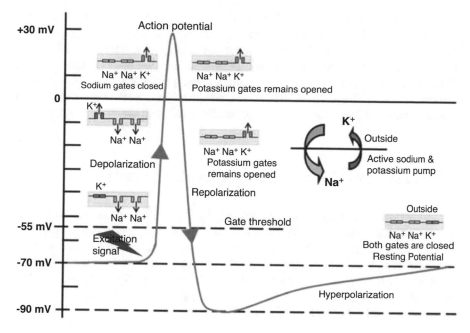

Fig. 1.1 Overview of the sodium (Na^+) and potassium (K^+) pump mechanisms

cell membrane to form a high concentration of sodium (Na^+) ion gradient. Due to the presence of excess sodium (Na^+) (positively charged ions) outside the cell membrane, and then the negatively charged ions along the inner surface of the cell membrane, the cell will be in a "resting state" or "polarized condition." Uneven charge distribution is caused by certain electrochemical reactions and processes in biological cells. The phenomenon of measuring potential is called "resting potential" (about -60 to -100 mV) [2].

When a specific cell is stimulated by providing ionic current or external stimulation or any trigger input, the characteristics of the cell membrane will be affected/changed with respect to the corresponding trigger. After the trigger input, the cell membrane begins to allow sodium (Na^+) ions to enter or flow into the cell. This effect is called the "avalanche effect." As more sodium (Na^+) ions flood into the cell, potassium (K^+) ions will start to slowly leave the cell. Because when the cell is at rest, the potassium (K^+) ion concentration gradient inside the cell is much higher. Now, the sodium (Na^+) ions (positive charge) in the cell membrane will cause the polarity of the cell membrane to change. As a result, the equilibrium conditions of the cell's previous resting state are disrupted, leading to a measurable potential change across the cell membrane, which can be called an "action potential" (usually in the +20-mV peak range). The cell will be in a depolarized state. The outer side of the cell membrane is temporarily negative with respect to the inner cell membrane [3].

After a period of time (usually 1 ms), by allowing potassium (K^+) ions to flow into the cell, the cell is polarized again, thereby returning to a normal depolarization/ quiescent state. Therefore, the cell will return to a resting state. It is a "repolarization" process, similar or the same as depolarization; depolarization involves sodium (Na^+) ions as the main ion, whereas in repolarization, potassium (K^+) ions play an important role. This is the origin of biosignals or bioelectrical signals or biopotentials. All biomedical signals we measure are based on this mechanism.

1.3   Characteristics of General Biomedical Signals

As aforementioned, cells, tissues, and/or organs, individually or in groups, continuously generate some measurable voltage/current signals that are transmitted to various parts of the body through our neural network to maintain the normal life and activities of the organism (e.g., heartbeat and breathing). These bioelectrical signals are usually time-varying signals, which can be easily described by their amplitude and frequency, and sometimes they can also be described by phase. Some common biomedical signals are listed below:

1.3.1   Electroencephalogram (EEG)

The origin of this signal is the movement of ionic currents between neuronal cells (i.e., the flow of ions from the postsynaptic junction of one neuronal cell to the next presynaptic junction of other neuronal cells connected to it, or just stimulate neurons), and it exists in the human brain. Ionic current accumulates in the cerebral cortex and reaches the scalp as an EEG signal through the skull, which can be recorded by placing electrodes on the scalp.

Neuromuscular diseases, such as amyotrophic lateral sclerosis (ALS), myasthenia gravis, multiple sclerosis, and spinal muscular atrophy, are some serious diseases caused by the progressive degeneration of nerve cells in the spinal cord and brain. This disease leads to a lack of control over the coordination of various parts of the body. The instrument used to record EEG is called "electroencephalograph."

- Electrooculogram (EOG)

  The source of this signal is the electrical potential (due to eye movement) generated in the corneal retinal muscles. By placing (surface) electrodes on the horizontal and vertical axes on both sides of the human eye, the reference electrode is located at the center of the forehead, and information about eye movement or rotation can be measured. The instrument used to record EOG is called "electrooculography."

- Electrocardiograph (ECG)

    The source of this signal is electrical activity related to heart function. Electrical signals are usually generated from the atrial (SA) node (located in the right atrium (RA)) to stimulate atrial contraction. These signals then move to the atrioventricular (AV) node (located in the atrial septum). Afterward, these signals are divided into multiple signals, passing through the (left and right) His bundles to the respective Purkinje fibers (on both sides of the heart) and endocardium (on the apex) and to the ventricular epicardium, which causes the heart wall to contract and relax. By placing (surface) electrodes, due to the depolarization of the myocardium, a tiny potential change on the surface of the human body (skin) can be obtained with each heartbeat. The instrument used to record the electrocardiogram is called an "electrocardiograph."
- Electromyography (EMG)

    The origin of these signals is the electrical activity associated with the contraction and relaxation of muscle fibers. By placing electrodes, tiny potential changes can be obtained on the surface of the human body, but the EMG signal generated on the surface of the muscle will have more than one electrically activated muscle unit. This is the main disadvantage of the EMG signal. It is easily contaminated by signals. However, the EMG signals are widely used to understand the physiological state/characteristics of motor neurons and muscles; they also have various applications in biomedical engineering.

Table 1.1 lists some of the more common biomedical signals and their characteristics [4]. However, many physiological signals are not as easy as a doctor using a stethoscope to understand the heart rate and breathing status in the clinic. If we want to capture these weak biomedical signals, electrical engineers need to design amplifiers with different magnifications according to the different amplitudes of the biomedical signals, the frequency range of the filter, and the sampling rate of the ADC.

1.4 Biosignal Propagation

As aforementioned, because the time-varying action potential is generated across the cell membrane, it is transmitted to the outside through the neural network to form a spatial transmission to the relevant control unit. Using the experimental arrangement presented in Fig. 1.2, the spatial displacement of these action potentials can be observed. The nerve cells or axons are stimulated with long processes at one location, and their transmembrane voltage is measured at three different distances from the stimulation point. Each of the three measurement points recorded an action potential, but the time when the action potential occurred was proportional to the distance from the stimulus location. Therefore, the action potential propagates in space, and the propagation speed is limited, which is usually called nerve conduction velocity. The normal speed range is approximately 10–120 m/s [5].

Table 1.1 Biomedical signals [4]

Classification	Acquisition	Frequency range	Dynamic range	Comments
Bioelectric	Microelectrodes	100 Hz–2 kHz	10 µV–100 mV	
Action potential				Invasive measurement of cell membrane potential
Electroneurogram (ENG)	Needle electrode	100 Hz–1 kHz	5 µV–10 mV	Potential of a nerve bundle
Electroretinogram (ERG)	Microelectrode	0.2–200 Hz	0.5 µV–1 mV	Evoked flash potential
Electrooculogram (EOG)	Surface electrodes	dc–100 Hz	10 µV–5 mV	Steady-corneal-retinal potential
Electroencephalogram (EEG) Surface	Surface electrodes	0.5–100 Hz	2–100 µV	Multichannel (6–32) scalp potential
Delta range		0.5–4 Hz		Young children, deep sleep and pathologies
Theta range		4–8 Hz		Temporal and central areas during alert states
Alpha range		8–13 Hz		Awake, relaxed, closed eyes
Beta range		13–22 Hz		
Sleep spindles		6–15 Hz	50–100 µV	Bursts of about 0.2–0.6 s
K-complexes		12–14 Hz	100–200 µV	Bursts during moderate and deep sleep
Evoked potentials (EP)	Surface electrodes		0.1–20 µV	Response of brain potential to stimulus
Visual (VEP)		1–300 Hz	1–20 µV	Occipital lobe recordings, 200-ms duration
Somatosensory (SEP)		2 Hz–3 kHz		Sensory cortex
Auditory (AEP)		100 Hz–3 kHz	0.5–10 µV	Vertex recordings
Electrocorticogram	Needle electrodes	100 Hz–5 kHz		Recordings from exposed surface of brain
Electromyography (EMG) Single-fiber (SFEMG)	Needle electrode	500 Hz–10 kHz	1–10 µV	Action potentials from single muscle fiber
Motor unit action potential (MUAP)	Needle electrode	5 Hz–10 kHz	100 µV–2 mV	
Surface EMG (SEMG)	Surface electrodes			
Skeletal muscle		2–500 Hz	50 µV–5 mV	
Smooth muscle		0.01–1 Hz		
Electrocardiogram (ECG)	Surface electrodes	0.05–100 Hz	1–10 mV	

(continued)

Table 1.1   (continued)

Classification	Acquisition	Frequency range	Dynamic range	Comments
High-frequency ECG	Surface electrodes	100 Hz–1 kHz	100 μV–2 mV	Notchs and slus waveforms superimposed on the ECG

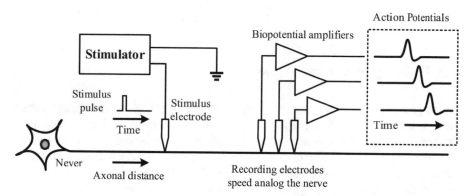

Fig. 1.2   Use spatially distributed electrodes to record propagated action potentials

1.5   Biomedical Signal Measurement

The measurement requires a system-based model to reasonably interpret the measurement results. For example, when we measure the blood pressure, we treat the heart as an oscillating pump, which causes periodic changes in the pressure of the cardiovascular system. Conversely, when we measure the air pressure in the tire or the water pressure in the water supply pipe, if we see the pressure change over time, it means that there is a problem.

The physiological measurement principle block diagram in Fig. 1.3 presents signal pickup, followed by analog processing and output. Signals are produced by physiological processes, usually physical quantities that change over time or space. The transducer converts this physical quantity to an electrical signal suitable for subsequent processing by the instrument. Analog processing includes amplifiers used to amplify the desired signals and filters used to suppress unwanted "noise." The output device is a display or paper chart recorder used to present information to the user. Most modern instruments convert analog signals to digital forms suitable for computer analysis. The digitized signal can be analyzed on the computer immediately when the signal enters, or it can be stored in the computer for a more complex analysis later. If it is an implantable system, the DSP processing or biomedical signal on the implanted chip is transmitted to an external receiving device via radio frequency for subsequent signal processing and recording.

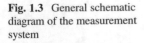
Fig. 1.3 General schematic diagram of the measurement system

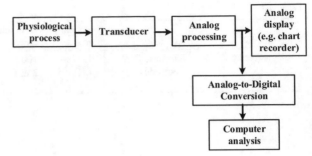

1.5.1 Interference and Noise

In the process of recording these biological signals, some unwanted signals, called noises, are also picked up. These noises may be inherent to the measurement equipment or may be generated by other systems in the vicinity of the recording. In physiological measurement, these unwanted noises will have a significant impact on their analysis and interpretation. For example, (a) when using the chest lead to record the electrocardiogram, the EMG of the intercostal muscles can be obtained at the same time; (b) when recording the EMG of the back muscles, the EMG can also be obtained at the same time. In the first case, ECG is the desired signal, and EMG is the unwanted signal, which we call "noise." In the second case, EMG is the desired signal, and ECG is the unwanted "noise" signal. In both cases, unwanted noise signals are unavoidable as they are electrical events that occur in the body. Choosing a noise removal technique requires not only an understanding of the characteristics of noise but also a clear understanding of physiology. Use a common example to illustrate the relationship between noise and signal. For drivers, traffic sounds may contain useful information, and telephone sounds may interfere with driving; however, for office workers, traffic noise may be an unwanted disturbance. Telephone conversation is a useful signal.

Noise and signal usually mix well, but it is difficult to separate them. Converting the measurement signal to another domain or space can better separate signal and noise. The more separated or different they are, the easier it is to remove the noise; otherwise, some signals may also be discarded along with the noise. The concept of converting a signal to another domain is very useful. For example, to convert a signal from the time domain to the frequency domain, a filter can be used to obtain the desired signal.

Noise removal usually requires a trade-off between the amount of noise to be removed and the amount of signal to be retained. A full understanding of this trade-off is important to effectively reduce noise. The importance of this compromise cannot be overstated. In the past few decades, noise reduction algorithms and

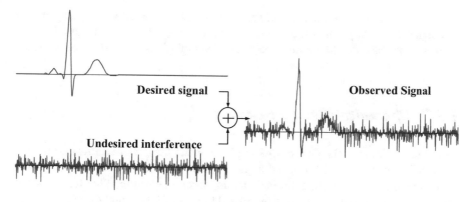

Fig. 1.4 Additive noise: random noise added to ECG

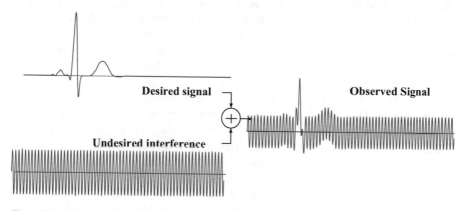

Fig. 1.5 Additional noise: 50/60-Hz power line interference added to ECG

technologies have made tremendous progress. Advances in electronic technology, new algorithms, and advances in computing power and speed have all contributed to great noise reduction. However, people should not be indifferent to noise pollution because of the idea that it can always be eliminated, because elimination is almost always imperfect.

Figure 1.4 presents an example of a randomly varying noise signal added to the desired signal (in this case, ECG). In the generated signal, it is difficult to distinguish the characteristics of the electrocardiogram. This kind of random noise is usually caused by thermal effects in electronic equipment [5].

Figure 1.5 presents an example of a rhythm noise signal [5], a sinusoidal signal added to the desired signal (ECG). Here, the characteristics of the electrocardiogram are also difficult to distinguish. In this case, the source of sinusoidal noise is electromagnetic interference from building power lines. In the case where all signals are contaminated by noise, reduction of noise pickup should be tried first by improving the measurement settings. For noise from external electromagnetic sources, the noise can be significantly reduced by using conductive shielding around the signal line. Only when the method of eliminating all the noise caused by the environment fails should the noise reduction technology be used after acquisition.

1.5.2   Biomedical Signal Processing

The core of biomedical signal processing is to obtain life system status information extracted from biological and physiological systems. Therefore, the monitoring and interpretation of this information have an important diagnostic value for clinicians and researchers to obtain information related to human health and diseases.

Biomedical signals are mainly used to diagnose or detect specific pathological or physiological conditions. In addition, these signals are also utilized to analyze medical prevention and healthcare research. When processing biomedical signals, due to the complexity of the underlying biological structure and its signals, it largely depends on the understanding of the signal source and many special properties. Because biomedical signals are usually measured via indirect, noninvasive measurements, the detected signal is usually destroyed by noise, and the relevant information is not easy to "see at a glance," and sometimes it is even difficult to extract from the original signal. Therefore, it is necessary to process the original signal; denoise the signal; accurately identify the signal model through analysis, feature extraction, and dimensionality reduction; clarify the information revealed by the signal; and use machine learning technology to predict future functions or pathological events. Because biomedical signals are usually well localized in time (such as spikes) and other diffuse feature (such as small oscillation) combination [6], advanced signal processing and machine learning techniques can help identify these features for correct judgment and processing.

The biomedical signals collected from the body can be at the organ level, cell level, or molecular level. There are several common biomedical signals, such as the signal presented in electrocncephalogram (EEG), which is the electrical potential related to the brain; electromyography (EMG), which is related to the muscle potential; and electroretinogram (ERG), which is the electrical potential related to the retina.

Biomedical signals contain numerous types of artifacts, including internal or external interference noise. Most artifacts and noise can be filtered out by using signal denoising techniques. Biomedical signal analysis and processing are mainly divided into three steps: preprocessing/denoising, feature extraction/dimensionality reduction, and classification/classification, as presented in Fig. 1.6 [7].

Preprocessing/denoising: The main goal of preprocessing/denoising is to simplify subsequent procedures without losing relevant information and to improve signal quality by increasing the signal-to-noise ratio (SNR). Filters and transformations are often used in the preprocessing procedure. Researchers use these methods to eliminate or at least reduce unwanted signal components by converting the signal.

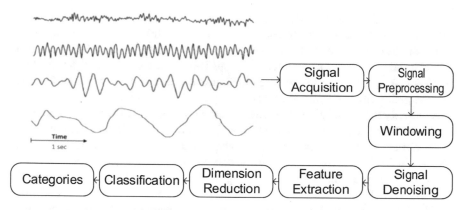

Fig. 1.6 General biomedical signal analysis framework

Feature extraction: Biomedical signals are composed of multiple data points, and different feature extraction methods can be used to extract information features. These unique and informative parameters describe the behavior of signal waveforms that specify precise actions. The biomedical signal pattern can be expressed in terms of frequency and amplitude. Different function extraction algorithms can be employed to extract these functions. Biomedical signals can also be decomposed using time–frequency (TF) methods, which can detect changes in time and frequency. Dimensionality reduction is the process of reducing the original feature vector while retaining the most distinctive information and removing irrelevant information [8].

Signal classification: The framework analyzes the salient features of the signal and determines the signal category based on these salient features.

1.5.3 Biomedical Signal Recognition

1.5.3.1 Electroencephalogram (EEG)

The brain is the most important organ of the human body. The electrical signals generated represent the function of the brain and the state of the entire body. Therefore, the electroencephalogram (EEG) signal obtained from the brain should also correspond to the state of the subject's entire body. The neurophysiological characteristics of the brain and the correlation of the generated signals need to be understood to obtain the basic working principles, and signal processing techniques should be used to identify, analyze, and treat brain diseases. EEG provides a method for diagnosing various diseases and abnormalities of the human nervous system.

An electroencephalogram (EEG) is a record of the electrical potentials produced by the brain. Analysis of the frequency components of the EEG signals usually considers the following frequency bands: delta wave (up to 4 Hz), theta wave

(4–8 Hz), alpha wave (8–12 Hz), beta wave (12–26 Hz), and gamma wave (26–100 Hz). The activities of each wave will have different reactions according to the differences related to the mental and physical tasks. The duration of the EEG signal is 0.01–2 s, and the amplitude is about 100 μV and usually less than 300 μV. EEG is mainly performed in research institutions and medical institutions to detect disease pathology and epilepsy. Seizures represent unexpected bursts of wild electrical activity in a population of neurons in the cerebral cortex. These fluctuations can cause irregular neuron synchronization and seizures, which can affect consciousness, sensation, or movement. Electroencephalogram (EEG) signals are usually checked using spectrum (α, β, θ, δ, γ) analysis methods. Once a transient event such as epilepsy occurs, a series of unexpected events or spikes will appear in the recorded signal.

Manual interpretation of the EEG is very time-consuming, as these records may last for hours or days. This is also an expensive process as it requires well-trained experts. Therefore, the use of machine learning to assist in the interpretation of EEG can help reduce the burden on medical staff. Machine learning optimizes parameters by executing learning algorithms and using training data to find optimal parameters. These training data are historical data or previous experience. The main learning objective can be predictive to make predictions from labeled data, such as classification and regression models. Machine learning tools can also classify patterns in EEG to enhance recognition capabilities, especially in biomedical signal analysis. They can complete real-time identification and diagnosis.

1.5.3.2   Electrocardiograph (ECG)

The human heart is the most important muscle of the human body. It forms the cardiovascular system with blood vessels and pumps blood into every cell of the body. An electrocardiograph (ECG) is a noninvasive instrument that captures the electrical activity of the heart and displays irregular heartbeats. Therefore, the electrocardiograph is an important tool for judging the health and function of the cardiovascular system.

The electrocardiograph (ECG) signal is derived from the electrical activity on the body surface of the heart, so the equipotential surface can be calculated and analyzed over time. Modern electrocardiograph equipment fully combines an analog front end, an analog-to-digital (A/D) converter, a dedicated input and output (I/O) processor, and a microcomputer.

A potential produced by electrical activity is related to heart function. Heart function is obtained by stimulating the electrical pulses (repetitive periodic synchronous rhythm signals) generated by the sinus (SA) node (located on the right) that controls the atrium (RA) by stimulating atrial contraction. These signals then move to the atrioventricular (AV) node (located in the atrial septum). Subsequently, the signal is divided into multiple signals, passing through the (left and right) His bundles to the respective Purkinje fibers (located on both sides of the heart) and

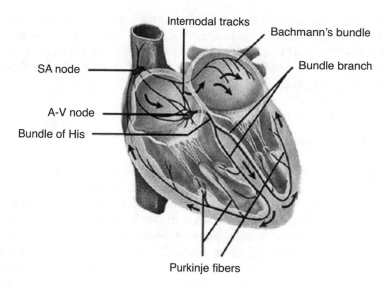

Internodal tracks

Bachmann's bundle

SA node

Bundle branch

A-V node

Bundle of His

Purkinje fibers

Fig. 1.7   Cardiac conduction system

Fig. 1.8   Typical ECG

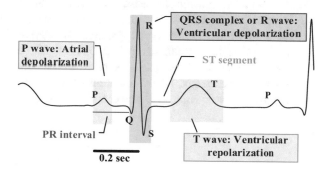

P wave: Atrial depolarization

QRS complex or R wave: Ventricular depolarization

ST segment

PR interval

0.2 sec

T wave: Ventricular repolarization

endocardium (located on the apex), reaching the ventricular epicardium, causing the heart wall to contract and relax, as presented in Fig. 1.7 [9].

Figure 1.8 presents a typical ECG signal [10]. ECG signals are usually in the 2-mV peak-to-peak range and occupy a bandwidth of 0.05–150 Hz. The characteristics of the body surface wave depend on the number and relative speed of the activated tissue at a time, the direction of the activation waveform (action potential), and the position of the electrode. The first ECG wave of the cardiac cycle is the P wave, which represents atrial depolarization. The conduction of the heart pulse is carried out from the atria through a series of specialized heart cells (the atrioventricular node and the hippo system). After the P wave, there is a relatively short period of time when the potentials are relatively equal. This is the P-Q interval defined by the propagation delay time of specialized cells, usually 0.2 s. Once the large muscles of the ventricle are stimulated, there will be rapid and large deflection on the body surface. The depolarized waveform of this ventricle is often called the

QRS complex. After the QRS complex is another equipotential segment, namely, the S-T interval. The S-T interval represents the duration of the action potential, usually around 0.25–0.35 s. After this short period of time, the ventricles return to their electrical resting state, and the repolarization wave is considered a low-frequency signal called T wave [9].

1.5.3.3 Electromyography (EMG)

The responsibility of the human musculoskeletal system is to obtain the power needed to perform various activities. The system is composed of the nervous system and the muscular system, which together constitute the neuromuscular system. The movement and alignment of the limbs are controlled by electrical signals transmitted back and forth between the muscles and the peripheral and central nervous systems. Neural information is encoded as an action potential rate (discharge rate) that is proportional to the stimulus. The signal carried by the nerve is a series of frequency-encoded pulses, which are action potentials. For all neural information transmission, whether it is to transmit the required intensity information to the motor neurons of the muscles or to transmit the sensory intensity to the sensory neurons of the brain, the information contained is in the form of action potentials. The required force determined by the central nervous system is encoded as the frequency of the action potential, and at the muscle, this information is decoded as the corresponding force. The force generated by the muscle increases with the increase in the action potential or stimulation frequency. The stimulation frequency is the reciprocal of the stimulation pulse interval. Figure 1.9 presents a typical force–frequency curve [6].

Fig. 1.9 Muscle strength-frequency curve

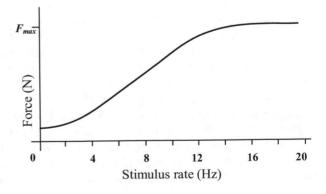

1.6 Application of Biosignals

1.6.1 In the Field of Rehabilitation Engineering

Rehabilitation engineering involves the use of systematic applications of engineering principles to improve the quality of life (QOL) of individuals with multiple levels of disabilities. In the past, researchers have developed assistive devices, such as electric wheelchairs for the disabled, smart canes for the visually impaired, hearing aids, manipulators or bionic arms for amputees, etc., to improve the quality of life in any way. Rehabilitation assistive devices are mainly used to help the disabled, so they must have accurate, fully functional, and easy-to-use assistive devices/equipment. In addition, rehabilitation aids driven by biological signals combine human biological signals with computers and machines, making the auxiliary equipment/equipment an extension of the human body. Human–computer interaction/human–machine interaction (HCI/HMI) involves connecting the human biological system to a rule-driven intelligent computer to command/control the machine to perform specific tasks. We have listed examples of each type of biosignal-driven rehabilitation aids.

1.6.1.1 Rehabilitation Aid Driven by the EEG Signal

EEG is produced by the movement of ionic currents between neuronal cells in the brain. It passes through the skull and reaches the surface of the scalp with an amplitude and frequency of approximately 2–200 µV and 0.1–100 Hz, respectively. Therefore, by placing appropriate surface electrodes on different positions of the scalp (as presented in Fig. 1.10), EEG signals can be recorded according to the desired wave type. Generally, researchers and clinicians use *Ag/AgCl* (or gold-plated) surface electrodes to acquire EEG signals and avoid fixation problems. The collected original EEG signal is very small, and a certain degree of amplification is performed with an appropriate amplification level. Subsequently, appropriate filtering techniques need to be used to eliminate unwanted signals. Through ADC/DSP, the decoded EEG signal is fed to the controller. The controller will make decisions based on a predefined set of rules and send corresponding commands to the machine to perform specific tasks, for example, controlling the connected electric wheelchair or robotic arm according to the EEG signals [11].

1.6.1.2 Rehabilitation Aid Driven by the EOG Signal

The electrical potential generated by the movement of the eyeball in the cornea and retinal muscles is called the EOG signal. The frequency range of the EOG signal is about 0.1–10 Hz, and the peak amplitude is 0.04–3.5 mV. Therefore, measurements can be made by placing surface electrodes on the horizontal and vertical axes on both

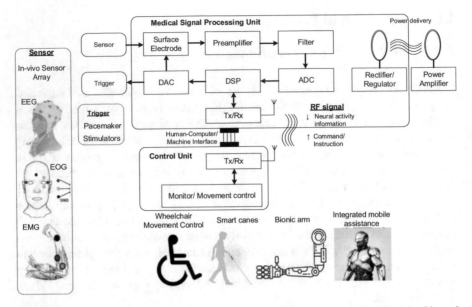

Fig. 1.10 A medical signal processing unit can be applied to various rehabilitation aids and implanted devices

sides of the human eye and placing the reference electrode in the center of the forehead or elsewhere, as presented in Fig. 1.10. To avoid loss of information, both horizontal and vertical signals use pre-amplification stages. Subsequently, to prevent unwanted signals, appropriate filtering techniques need to be used. Through the interface channel, the filtered EOG signal is fed to the controller. The controller sends the filtered EOG signal to undergo the feature extraction process in order to extract useful information and then makes a decision based on this information considering predefined rules; subsequently, the controller sends the corresponding commands to the machine to perform specific tasks. For example, if an electric wheelchair is connected to the controller, the movement of the electric wheelchair will be determined by considering predefined rules, and the corresponding commands will be transmitted to the machine to perform specific tasks. If the electric wheelchair is connected to the controller, the sports wheelchair will rely on the EOG signal generated by the calculation method and the corresponding control signal and then carry instructions to control auxiliary equipment [12].

1.6.1.3   Rehabilitation Aid Driven by the EMG Signal

The EMG signal is generated by the electrical activity related to the contraction and relaxation of muscle fibers. Its frequency range is about 5–3000 Hz, and its amplitude is about 0.1–10 mV. Therefore, by placing the surface electrode on the corresponding muscle, the potential change exhibited by the muscle surface can be obtained. Therefore, the acquired original EMG signal will be pre-amplified and filtered, respectively. The filtered EMG signal is fed to the ADC/DSP and then connected to the controller of the auxiliary equipment. Then, predefined rules are made based on the information to perform specific tasks, as presented in Fig. 1.10 [13].

1.6.2   In the Field of Implantable Medical Devices

Like rehabilitation engineering, implantable medical equipment can also sense human biomedical signals, thereby restoring bodily functions and improving the quality of life. However, unlike rehabilitation engineering, it is implanted in the body, like a pacemaker [14, 15], to constantly observe the electrical movement of the heart. When it detects an arrhythmia, it applies electrical stimulation (through electrodes) to the heart to correct its heart rate, help the heartbeat, and directly save lives. Or we can directly perceive sounds through cochlear implants [16, 17] using electronic devices to recognize and encode sounds and then stimulate the auditory nerve to enable deaf people to hear. These two implanted devices have restored function to tens of thousands of patients.

With the advancement of science and technology, in addition to gaining a certain understanding of the generation and transmission of biomedical signals, the research community has also exhibited great interest in the subject of neuroengineering as it can be used to treat diseases. Neural engineering is the use of engineering techniques to understand, improve, reconstruct, enhance, or otherwise utilize the characteristics of the nervous system. Neuroscience in advanced systems relies on the ability to monitor and report electrical signals generated by neurons in the brain or external nervous system. Simultaneous recording of small signals from brain nerve potentials on a large number of electrodes provides an effective method for neuroscientists and clinicians to study the dynamic changes of the brain state and understand various neurophysiological and behavioral characteristics. The implantable biomedical system combined with wireless technology can not only control the peripheral equipment and assist the functions of the human body like the rehabilitation engineering, as the patient's movement will not be restricted due to the wireless technology. It can also do more than the rehabilitation engineering assistance to the human body, such as the detection and recovery of spinal cord injury, stroke, epilepsy, and Parkinson's disease. Experts estimate that 8–10% of Americans (approximately 20–25 million people) [18] or about 1 in 17 people in industrialized countries [19] carry some form of implanted devices.

1.7   Energy of Implanted Devices

Battery life is a key issue for all portable electronic devices. When portable devices become therapeutic implants, things become even more important. The life of these devices may depend on battery life. Generally, the implanted battery must be replaced after a certain period of time, which is no different from all battery-powered devices. Since the patient will undergo the necessary surgery to replace the implanted battery, it is not expected to be replaced frequently. This encourages researchers to develop dynamic solutions for implants [20–22]. Regardless of the power source of the implant (e.g., battery, piezoelectric power, inductive link, or a combination of these sources), reducing the power consumption of the implanted circuit and using ultra-low-power implanted devices are the most important. Reducing the power consumption in these circuits also helps reduce the chance of damage to surrounding tissues due to heat dissipation [23]. Therefore, to prolong the life of the implanted device and avoid harm to the human body, the power consumption of the device itself is the most important indicator.

1.8   Wireless Frequency Band of the Implanted Device

As presented in Fig. 1.10, the biomedical signal is amplified, filtered, and converted to a digital signal by ADC, and then the required signal characteristics are captured by the DSP. The collected biomedical signals or the parameter settings of the implanted device to be changed can be transmitted through the existing wireless frequency band, as presented in Table 1.2 [24]. Since the frequency and power of the wireless transmitter and receiver determine the life and size of the device, the choice of wireless frequency band is very important for implantable biomedical chips with extremely low power consumption. Since WLAN uses the 2.4-GHz frequency band for multiple access and transmits greater signal power, many other wireless systems and devices that use the 2.4-GHz frequency band, such as Zigbee (IEEE 802.15.4) and Bluetooth (IEEE 802.15.1), may be strongly interfered with by WLAN [25]. UWB is not suitable for implantable chips. Not only is the power consumption of UWB not as low as the literature [26] claims, but the high operating frequency (3–10 GHz for medical applications) makes UWB a failure with high penetration rates. MICS (Medical Implant Communication System) [27] frequency band is used to collect sensor signals, with low transmission power (25 Ω, equivalent to UWB) and low power consumption; it also provides the most suitable frequency band for implantable chips [28].

In addition to being used at home or in the workplace, MICS can also be combined with WMTS in a multi-hop structure for wireless telemetry in hospitals [8], as presented in Fig. 1.11. Both frequency bands are internationally available, allowing multiple patients to be monitored simultaneously. In the communication process, it is necessary to use a phase-locked loop (PLL) or a frequency synthesizer

Table 1.2 Existing wireless frequency band [8]

	MICS	WMTS	UWB IEEE (802.15.6)	IEEE 802.15.4 (Zigbee)	IEEE 15.1 (Bluetooth)	WLANs (802.11b/g)
Frequency bands	402–405 MHz	608–614 MHz, 1395–1400 MHz, 1429–1432 MHz	3–10 GHz	2.4 GHz, (868/915 MHz Eur./US)	2.4 GHz	2.4GHz
Bandwidth	3 MHz	6 MHz	>500 MHz	5 MHz	1 MHz	20 MHz
Data rate	16 kbps (AMIS)	76 kbps	850 kbps	250 kbps(2.4 GHz)	1 Mbps	>11 Mbps
Multiple access	CSMA/CA, Polling	CSMA/CA, Polling	ALOHA	CSMA/CA	FHSS	OFDMA, CSMA/CA
Tx power	−16 dBm (25 μW)	≥10 dBm and <1.8 dB (1.5 watt)	41 dBm	0 dBm	4 dBm, 20 dBm	250 mW
Range	0–10 m	>100 m	2 m	0–10 m	10–100 m	0–100 m

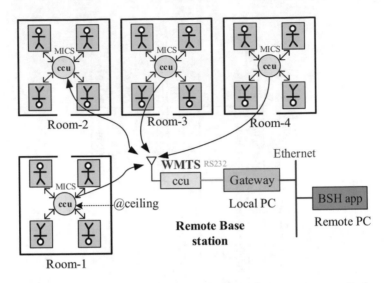

Fig. 1.11 A wireless body area network system for multi-patient monitoring in medical centers [8]

to meet the specifications of the wireless frequency band and provide a high-quality
oscillation frequency. From the above discussion, we can know that in the biomedical signal processing unit, in addition to DSP, ADC/DAC and frequency synthesizer
are two important components that are indispensable for most implanted devices.

1.9   Organization

This book focuses on the design and implementation of implantable frequency
synthesizers and analog-to-digital converters under ultra-low voltage and low
power consumption. The next few chapters are organized as follows:

Chapter 2 Low-Power and Low-Voltage VLSI Circuit Design Techniques for
Biomedical Applications

Chapter 3 Introduction to Frequency Synthesizer

Chapter 4 Implementation and Theoretical Analysis of Low-Power 0.35-V Frequency Synthesizer for Implanted Systems

Chapter 5 Introduction to ADC

Chapter 6 Implementation and Theoretical Analysis of Low-Power 0.3-V ADC
for Implanted Systems or Environmental Sensing

Chapter 7 Conclusions

References

1. Webster, J. G. (Ed.). (2009). *Medical instrumentation: Application and design*. Wiley.
2. da Silva, H. P., Fred, A., & Martins, R. (2014). Biosignals for everyone. *IEEE Pervasive Computing, 13*(4), 64–71.
3. Dimitrova, N. A., & Dimitrov, G. V. (2003). Interpretation of EMG changes with fatigue: Facts, pitfalls, and fallacies. *Journal of Electromyography and Kinesiology, 13*(1), 13–36.
4. Bronzino, Joseph D. "The biomedical engineering handbook: Medical device and systems, third edition," 2006.
5. Devasahayam, S. R. (2012). *Signals and systems in biomedical engineering: Signal processing and physiological systems modeling*. Springer.
6. Unser, M., & Aldroubi, A. (1996). A review of wavelets in biomedical applications. *Proceedings of the IEEE, 84*(4), 626–638.
7. Saikia, A., & Paul, S. (2019). EEG signal processing and its classification for rehabilitation device control. In *Application of biomedical engineering in neuroscience* (pp. 173–196). Springer.
8. Wołczowski, A., & Zdunek, R. (2017). Electromyography and mechanomyography signal recognition: Experimental analysis using multi-way array decomposition methods. *Biocybernetics and Biomedical Engineering, 37*(1), 103–113.
9. Sutton, R., & Bourgeois, I. (1991). *The foundations of cardiac pacing. Part I*. Futura Publishing Company.
10. Bronzino, J. D. (2000). *The biomedical engineering handbook* (Vol. 1, 2nd ed.). CRC Press and IEEE Press.
11. Garrett, D., et al. (2003). Comparison of linear, nonlinear, and feature selection methods for EEG signal classification. *IEEE Transactions on Neural Systems and Rehabilitation Engineering, 11*(2), 141–144.
12. Ferreira, A., et al. (2007). Human–machine interface based on muscular and brain signals applied to a robotic wheelchair. *Journal of Physics: Conference Series, 90*(1), 012094.
13. Uvanesh, K., Nayak, S. K., Champaty, B., Thakur, G., Mohapatra, B., Tibarewala, D. et al. (2016). Classification of surface electromyogram signals acquired from the forearm of a healthy volunteer. In *Classification and clustering in biomedical signal processing*, IGI Global, Hershey, pp. 315–333.
14. Ward, S. H. C., & Metcalfe, N. H. (2013). A short history on pacemakers. *International Journal of Cardiology, 169*(4), 244–248.
15. Vardas, P. E., Simantirakis, E. N., & Kanoupakis, E. M. (2013). New developments in cardiac pacemakers. *Circulation, 127*(23), 2343–2350.
16. Zeng, F.-G., Rebscher, S., Harrison, W., Sun, X., & Feng, H. (2008). Cochlear implants: System design, integration, and evaluation. *IEEE Reviews in Biomedical Engineering, 1*, 115–142.
17. Wilson, B. S., & Dorman, M. F. (2008). Cochlear implants: A remarkable past and a brilliant future. *Hearing Research, 242*(1–2), 3–21.
18. (2000) .FDA Consumer. *FDA Consumer, 34*(2), 7.
19. Thwaites, T. (1995). Total recall for medical implants. *New Scientist, 145*(1968), 12–13.
20. Simard, G., Sawan, M., & Massicotte, D. (2010). High-speed OQPSK and efficient power transfer through inductive link for biomedical implants. *IEEE Transactions on Biomedical Circuits and Systems, 4*(3), 192–200.
21. Chen, H., et al. (2009). Low-power circuits for the bidirectional wireless monitoring system of the orthopedic implants. *IEEE Transactions on Biomedical Circuits and Systems, 3*(6), 437–443.
22. Platt, S. R., Farritor, S., Garvin, K., & Haider, H. (2005). The use of piezoelectric ceramics for electric power generation within orthopedic implants. *IEEE/ASME Transactions on Mechatronics, 10*(4), 455–461.

23. Sarpeshkar, R., et al. (2008). Low-power circuits for brain–machine interfaces. *IEEE Transactions on Biomedical Circuits and Systems, 2*(3), 173–183.
24. Yuce, M. R., & Ho, C. K. (2008). Implementation of body area networks based on MICS/WMTS medical bands for healthcare systems. In *30th Annual International Conference of the IEEE Engineering in Medicine and Biology Society*, pp. 3417–3421.
25. Shin, S. Y., Park, H. S., & Kwon, W. H. (2007). Mutual interference analysis of IEEE 802.15.4 and IEEE 802.11 b. *Computer Networks, 51*(12), 3338–3353.
26. Gharpurey, R., & Kinget, P. (2008). Ultra wideband: Circuits, transceivers and systems. In *Ultra Wideband* (pp. 1–23). Springer.
27. Chiao, J.-C. (2017). Wireless closed-loop stimulation systems for symptom management. In *Principles and Applications of RF/Microwave in Healthcare and Biosensing*, pp. 151–189.
28. Higgins, H. (2007). *In-body RF communication and the future of healthcare.* Zarlink semiconductoer.

Chapter 2
Low-Power and Low-Voltage VLSI Circuit Design Techniques for Biomedical Applications

In this chapter, we will discuss low-power and low-voltage VLSI circuit design. From the evolution of technology, we can see the changes and considerations and weigh the key points to achieve the best design performance. First, we will introduce the many benefits brought about by the continuous advancement of semiconductor technology, such as smaller areas, faster speed, more functions, and lower power consumption. However, the progress of semiconductor technology has also brought us some problems, such as various leakage currents. Next, we will discuss how to design low-voltage and low-power consumption in digital circuit design, including possible solutions and suggestions to reduce dynamic power consumption, static power consumption, and leakage current. Then, a trade-off is made between the performance of the digital circuit and the low-power consumption under voltage scaling. Next, we will introduce the problems faced by low-voltage analog circuit design and then discuss traditional design methods and G_M/I_D design methods. Finally, we will discuss some considerations for the design and implementation of nano-analog circuits.

2.1 Process Technology Evolution and Power-Supply Voltage Scaling

On the left illustration in Fig. 2.1, we see the scaling relationship between process technology evolution and power-supply voltage, which is mainly based on the typical power-supply voltage data of ITRS across technology nodes in the past few years. Let us look at the voltage scaling trend. In the past, it used to scale the voltage of each generation process by 30%. So if the design consumes too much power, then all we have to do is to scale it to fit the next process technology. Then, the new voltage will be 0.7 of the original voltage, and the power will be 0.7 times

© The Author(s), under exclusive license to Springer Nature Switzerland AG 2022
C.-C. Hung, S.-H. Wang, *Ultra-Low-Voltage Frequency Synthesizer
and Successive-Approximation Analog-to-Digital Converter for Biomedical
Applications*, Analog Circuits and Signal Processing,
https://doi.org/10.1007/978-3-030-88845-9_2

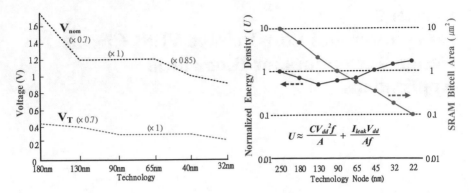

Fig. 2.1 Relationship between process technology evolution and power

0.7, which is only half of the original value. This is because the process has a built-in scaling mechanism that can reduce power consumption. In addition to the benefits of reducing the power number, scaling can also increase the operating frequency and also allow us to add more functions. However, in recent generations, the degree of voltage scaling has been significantly reduced, and now, we can see that the voltage is only reduced by 5–10% from one generation to the next, and the power gains from the process will be significantly reduced. Methods to reduce power consumption must change from process to circuit, including a large number of clock gating and voltage frequency adjustment, as well as many power management techniques. Moreover, we can see that the scaling of V_T is very slow because a lower V_T sometimes means more leakage. This is a trade-off between how much speed is gained and how much leakage will be lost. Thus, we see it stagnate at around 1.2–0.9 V at the 90-nm node.

The relationship between energy density and SRAM bit cell density and technology evolution is demonstrated by the right illustration in Fig. 2.1. In the gray curve, we can see that the SRAM bit cell density continues to increase as it continues to scale geometrically. In the red curve on the left axis, the energy density is used as an important indicator. The energy density equation is presented below, which is the sum of dynamic energy density and leakage energy density. Therefore, we divide the normal CV^2f energy by the area, as the area scaling continues; however, the voltage scaling is slow, and the frequency is also increasing, which means that the dynamic energy increases with technical capabilities. In the formula in the right illustration in the figure, the leakage power, which is the numerator, is divided by the area and frequency, because the leakage power is related to the operating frequency. Basically, the product of area and frequency also tends to decrease, and then, the energy density increases as the voltage becomes flat. We see an upward trend in the overall energy density. It means that we can continue to increase the number of transistors on the chip, but the power is so high that we cannot actually make full use of all these transistors. Some people call it "dark silicon."

2.2   Relationship Between Process Technology Evolution and Electrical Properties

There are three parameters used for manufacturing scaling: geometric scaling, V_{DD} voltage scaling, and threshold voltage scaling. The impact of technology scaling on circuit characteristics is a major concern for designers. Assuming that the geometric scaling ratio is s, the V_{DD} voltage scaling ratio and the threshold voltage scaling ratio are the same as p. The delay applied to a simple logic gate is estimated as follows:
The capacitive load is

$$C\frac{\tilde{\epsilon}_{ox}}{t_{ox} \cdot s}WL \cdot s^2\tilde{s} \tag{2.1}$$

The total capacitance is basically the scaling ratio of W over L is s. Therefore, the capacitance and transistor size are basically scaled by 0.7. The scaling ratio of V_{DD} is p.

$$V_{DD}\tilde{p} \tag{2.2}$$

The saturation of the current in the transistor (basically the switching current) is basically expressed by the following formula:

$$I_{sat}\tilde{\mu}C_{ox}\frac{W}{L}\left((V_{DD} - V_t)V_{dsat} - \frac{V_{dsat}^2}{2}\right)\tilde{p}; \tag{2.3}$$

It is consistent with the p ratio. The quadratic term is ignored here. Then, the delay time t_d is as follows [2]:

$$t_d\frac{C\tilde{V}_{DD}}{I_{sat}}\tilde{s} \tag{2.4}$$

t_d is indeed consistent with the scaling ratio s, which indicates that the delay time of a simple logic gate is shortened, or the performance of the circuit is improved with geometric scaling.

The geometric scaling ratio, as aforementioned, is usually set to half of the square root of 2, which is 0.7, and applied to W and L at the same time. Geometry has been scaling because size has always been the main driving force of Moore's Law. Starting at 180 nm, it performs constant field ($E = V/L$) scaling until we cannot maintain it. For 130 nm, with the development of technology, the supply voltage decreased to centered around 1.2 ~ 1 V. For reliability and heat dissipation reasons, it is indeed necessary to scale the voltage to a small amount, and the same is true for the threshold voltage. In fact, if the threshold voltage is too low, the transistor cannot be effectively turned off, and leakage occurs, so the threshold voltage scaling is basically limited by leakage.

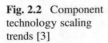
Fig. 2.2 Component technology scaling trends [3]

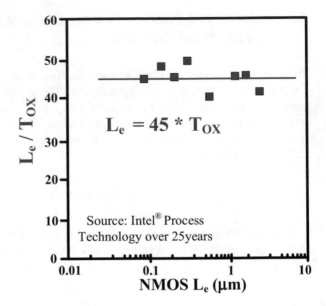

Figure 2.2 presents the trend scaling of device technology. On the x-axis, it is the L-effective value of the transistor; due to technological progress, the geometry of the transistor is scaled down to finer and finer geometries. On the y-axis, it is the ratio of the gate length of the transistor divided by the oxide thickness. We have noticed that these ratios have been very constant for about a few decades, indicating that whenever we scale to a smaller process geometry, the source and drain will get closer and closer. Therefore, the subthreshold leakage will increase intuitively as it is easier to jump through the channel. Also, because we are constantly scaling to fit finer geometries, every time we make the transistor gates smaller, we have to make the oxide thinner at roughly the same proportion. This means that the thinner the oxide, the easier it is for electrons to tunnel through it. The result is that transistors are scaled to finer geometries so that both subthreshold leakage and gate leakage are increasing.

2.3 Low-Voltage and Low-Power Digital Circuit Design

For digital circuits, the power is mainly composed of dynamic power and static power; there is a third smaller component called short-circuit power:

$$P = P_{\text{dynamic}} + P_{\text{static}} + P_{\text{short}} \tag{2.5}$$

For digital circuits, dynamic power is a function of the switching activity factor α multiplied by the sum of the capacitances of all nodes in the circuit C_L, the square of the voltage, and the operating frequency f:

$$P_{\text{dynamic}} = \alpha \cdot C_L \cdot V^2 \cdot f \qquad (2.6)$$

The static power is the power consumed when the circuit remains inactive, such as the energy consumption caused by leakage. The last short-circuit power, which is because the CMOS gate is turned on at the same time during the switching state, so the pull-up and pull-down transistors are both turned on for a very short period of time; then, one of them is turned off and the other is turned on, so some current will flow from the power supply to the ground during each switching operation.

2.3.1   Dynamic Power Reduction

Therefore, to reduce active power consumption, we start with the dynamic power formula and then try to process each element. First, reduce the first α, the factor of switching activities. It can be executed by conditional execution, conditional activation, conditional pre-charging, turning off blocks that are inactive for a period of time, and reducing long buses or high-capacitance nodes/buses. Because every time data is sent to the bus, the capacitors of the entire bus must be charged and discharged during the process. Therefore, if some parts of the bus are not used temporarily, they can be electrically isolated, and data can only be sent to the required parts without wasting power.

Second, by reducing the load capacitance C_L, more effective layout techniques can be employed to minimize the diffusion area of source and drain, wiring, and gate loads and focus on highly active nodes (such as clocks and dominoes). Another factor we can deal with is the voltage V^2. Thanks to squaring, anything we can do on voltage can get more benefits. Therefore, if the product runs at 1.0 V, we can scale it down (e.g., 0.7 and 0.8 V) to reduce power consumption. However, it is necessary to ensure that there is still an available application window on the voltage when designing, and scaling down will naturally slow down the operating speed. The last element is the clock frequency f. We can try to reduce it by using more parallelism and fewer pipeline stages or letting the flip-flop trigger on both the rising and falling edges, which is equivalent to doubling the speed. Therefore, the clock frequency is reduced by half, as well as the power consumption of the clock distribution network.

2.3.2 Dynamic Power Reduction Technology

Next, we will discuss common dynamic power reduction methods, namely, clock gating, multi-supply voltage design, and parallelism.

2.3.2.1 Clock Gating

Clock gating is the most popular active power reduction technique and has been widely used in most designs. The idea here is to save power by gating the clock during periods of very low data activity. As presented in Fig. 2.3, on the left, the enabled signal controls the MUX. The MUX continuously loops the old data back into the input, and the clock continuously reloads the value every cycle. Therefore, a lot of energy is wasted on the reloading of the same data on the flop. A better method is to enable the signal to gate the clock so that the latch will not be clocked until the new data is valid and waiting to be loaded. However, detailed timing analysis must be conducted on the design and simulation, as the enabled signal must arrive earlier and meet a certain setup time, and some logic verification is also required to ensure that the clock can be stopped and the state can be re-entered.

2.3.2.2 Multiple Power-Supply Voltages

Multiple power-supply voltages mean that we can have multiple power supplies within the same block or separate power supplies at each block level. Therefore, in a module, we must determine where the critical path is and maintain the maximum voltage on the critical path gate. Standard or lower voltage can be used anywhere on the noncritical path. The goal is to reduce power consumption as much as possible while keeping the system properly functioning. Of course, every time we change the voltage from low to high, we need a voltage level shifter. It occupies area and consumes power, so the challenge is to minimize the number of level shifters and define these voltage shifters to minimize the overhead of these voltage level shifters. As presented in Fig. 2.4, high-speed or critical paths use higher voltages, whereas other paths use lower voltages.

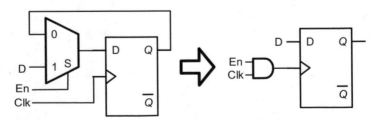

Fig. 2.3 Clock gating

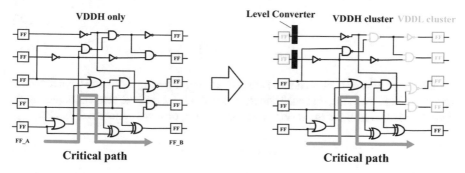

Fig. 2.4 Two-level supply voltage scheme [4]

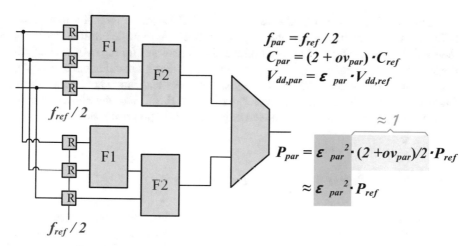

Fig. 2.5 Parallelism

2.3.2.3 Parallelism

Parallelism can be used as a way to improve performance [5], but from another perspective, any performance improvement technology or method can also be perceived as a way to reduce operating voltage but maintain operating performance and reduce overall dynamic power consumption. The original dynamic power consumption is as follows:

$$P_{\text{ref}} = C_{\text{ref}} V_{dd,\text{ref}}^2 f_{\text{ref}} \tag{2.7}$$

If we use the same design and duplicate it, as presented in Fig. 2.5, we perform two operations simultaneously. This increases the hardware cost because each function constructs two functional units that are replicated, but each path will run at half the frequency and then use MUX to ping-pong between them.

$$f_{\text{par}} = \frac{f_{\text{ref}}}{2} \qquad (2.8)$$

$$C_{\text{par}} = \left(2 + ov_{\text{par}}\right) \cdot C_{\text{ref}} \qquad (2.9)$$

Therefore, these functional units are time-division-multiplexed, the capacitance is approximately doubled, and each functional unit runs at half the reference frequency. This enables us to perform voltage scaling on each of these functional blocks.

$$V_{\text{dd,par}} = \varepsilon_{\text{par}} \cdot V_{\text{dd,ref}} \qquad (2.10)$$

$$P_{\text{par}} = \varepsilon_{\text{par}}^2 \cdot \left(2 + ov_{\text{par}}\right) \frac{P_{\text{ref}}}{2} \varepsilon_{\text{par}}^{\tilde{2}} \cdot P_{\text{ref}} \qquad (2.11)$$

There is twice as much hardware as the original hardware (actually more than twice the hardware) and the voltage drops by ε_{par}, a ratio to be determined, which is then substituted into Eq. (2.11). We find that it compensates for the overhead due to the squaring effect. For example, the paper [5] shows that it can either have twice the concurrency or save about 50% of battery power if the realization conditions are as follows: ov_{par} of 0.1, ε_{par} of 0.75, and P_{par} of $0.6 \times P_{\text{ref}}$. Under the allowable chip area, application parallelism can indeed reduce dynamic power consumption.

2.3.3   Static Power Reduction Technology

Here, several methods to reduce static power consumption will be introduced: power gating, MOSFET component leakage mechanism and its mitigation methods, the establishment of a low-leakage library, and different biasing techniques.

2.3.3.1   Power Gating

Power gating means to control the power supply to reduce the leakage current of two parameters related to V_{DD}.

$$P_{\text{leak}} = V_{\text{DD}} \cdot I_{\text{leak}} = V_{\text{DD}} \cdot I_0 \exp\left(\frac{-V_{\text{T}}}{nkT/q}\right) \qquad (2.12)$$

$$V_{\text{T}} = V_{\text{T0}} - \eta_{\text{DIBL}} V_{\text{DD}} \qquad (2.13)$$

Equation (2.12) directly reduces the power-supply voltage to reduce leakage power consumption. Equation (2.13) can obtain the benefit of reducing leakage from the relationship between DIBL and threshold voltage.

Figure 2.6 [6] presents the relationship between the supply voltage of 65-nm SRAM and the leakage current, where we can lower the supply voltage and reduce

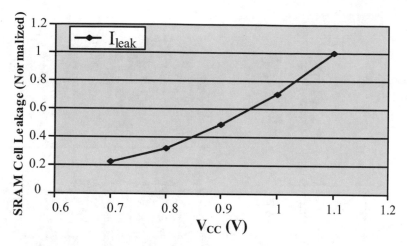

Fig. 2.6 Leakage reduction via supply voltage control

leakage. SRAM works under normal voltage (such as 1.1 V), but if we do not use it or access it temporarily, we can easily reduce the voltage to, for example, 0.7 V, which can reduce leakage by 80%.

2.3.3.2 Power Gating Applied to Memory (Sleep and Shut-Off Mode)

The other is about memory leak reduction technology [7], which can be achieved in two ways: one is to use PMOS sleep transistors, and the other is to use NMOS sleep transistors. Although both MOS sleep transistors can achieve similar functions, PMOS has better performance.

We will build a P-sleep transistor circuit on top of the memory sub-array. It has three operation modes. In the active mode, we turn on a large block of PMOS transistors, which will short-circuit the virtual VCC to the real VCC. The full voltage swing is available in the sub-array; thus, we read and write memory cells normally in this mode. The second mode is called sleep mode. In this mode, we only keep the content of memory; we do not read or write memory cells. Therefore, in this case, we will turn off the bulk of the PMOS transistors and keep them at that mode and then turn on the sleep bias transistors. This sleep bias transistor is adjustable and basically has four binary-sized branches. We can control the appropriate number of branches to reduce the virtual power supply from 1.1 V (namely, the nominal voltage) to about 850 mV. In terms of data retention, this is still a very sufficient voltage. However, we cannot read and write sub-arrays; we only maintain the content. Due to the lower voltage, our leakage is reduced by about three times, because 99% of the sub-arrays are idle—they just hold the value—and only 1% will be active, which is a good trade-off. The third mode is called "shut-off," which is used if we no longer need the contents of these sub-arrays or if we have permanently disabled these sub-arrays. Therefore, we turned off the shut-off transistor and then turned off all the branches of

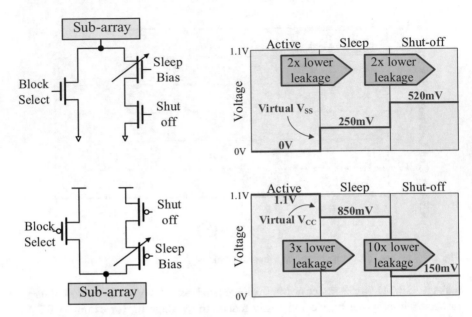

Fig. 2.7 Power gating applied to memory (sleep and shut-off mode)

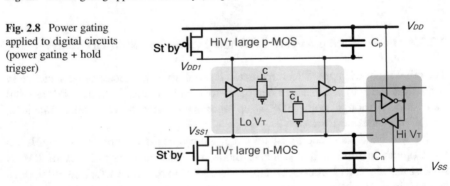

Fig. 2.8 Power gating applied to digital circuits (power gating + hold trigger)

sleep bias, so that the virtual power was completely reduced to about 150–200 mV, thereby reducing the leakage by ten times (Fig. 2.7).

2.3.3.3   Power Gating Applied to Digital Circuits (Power Gating + Hold Trigger)

If power gating is used, even if the power is cut off, the information may need to be stored in the designed flip-flop [8]. We need the so-called hold triggers (Hi V_T latch in Fig. 2.8). The idea is that there are two V_{DD}s on each flip-flop, one V_{DD} is always turned on to retain internal information, and the other V_{DD1}/V_{SS1} is switchable. When we turn off V_{DD1} in sleep mode, the information will be stored in the hold trigger.

When we turn on the device again, we can start running as it has the same information in the trigger.

We need to pay attention to some details in the actual implementation. First, the power gate or power gate unit is usually composed of small- and large-sized transistors, which prevent the power bounce from producing noise on the main V_{DD}. First, turn on the small gate, add a buffer, introduce a delay in the signal, and then turn on the larger transistor. Second, because we want the power switch itself to have as little leakage current as possible, high-threshold components are generally used for power switches.

2.3.4   Transistor Leakage Mechanisms and Solutions

2.3.4.1   Subthreshold Leakage I_{sub}

I_{sub}: Subthreshold leakage is the main leakage mechanism for processes beyond 90 nm, which may account for more than 90% of the leakage. When the gate voltage is lower than the threshold voltage, a small current flows from the source to the drain, thus escaping the gate control. This leakage is caused by carrier diffusion, so the inversion is very weak. When the transistor is turned off, this leakage flows from the drain to the source. Moreover, this leakage depends on V_{GS}, V_{DS}, and the length of the transistor (effective length):

$$I_{sub} = \mu_n C_{ox} \frac{W}{L} U_T^2 e^{1.8} e^{\frac{V_{GS}-V_T}{nU_T}} \left(1 - e^{\frac{-V_{DS}}{U_T}}\right) \tag{2.14}$$

$$U_T = \frac{KT}{q} \tag{2.15}$$

where V_{GS} denotes the gate-source voltage; V_T, the transistor threshold voltage; U_T, the thermal voltage; n, the subthreshold swing coefficient; μ_n, the surface mobility of electrons; C_{ox}, the capacitance per unit area of the gate electrode; W, the channel width; and L, the channel length. Therefore, when V_{GS} or V_{DS} increases, the leakage also increases, and it also increases with the increase in temperature. The solution at the process level is to use higher doping, shallower junction depth, thicker gate oxide, and good engineering design. Conversely, the solution at the design level is to use a larger L, transistor stacks, and multiple threshold voltage design or bias technique.

2.3.4.2   PN Junction Reverse Bias Current I_J

I_J: Reverse bias PN junction leakage. The PN junction reverse bias current is caused by the generation of electron–hole pairs due to minority carrier diffusion in depletion regions. For heavily doped source/drain, band-to-band tunneling (BTBT) current dominates. It goes from the drain to the body. This leakage will increase with V_{DD} and steep doping profile. There are always two leak solutions. At the process level, we need to perform good engineering design or silicon insulator (SOI) work. At the design level, bias technology is a good way to reduce the leakage.

Tunneling into and Through Gate Oxide I_G

I_G: Gate leakage, the gate leakage current caused by the nonzero quantum mechanical probability of charge tunneling through the gate oxide. The gate current affects the robustness and speed of the circuit. Processes beyond 90 nm have this problem. The leakage current will flow directly from the gate to the body. When the gate voltage is increased and the oxide is thinner, the gate leakage will be higher. For gate oxide thicknesses of the order of 2 nm and below, gate leakage becomes important. Because the thickness of silicon dioxide decreases by 2 Å, the leakage will increase by 10 times. To maintain C_{ox} and minimize leakage power, a thicker dielectric can be used, and the dielectric constant can be increased by changing the dielectric material. The latest process is to replace the gate silicon dioxide with a high-K dielectric. Since C_{ox} remains unchanged, the MOS transistors will maintain the same performance. C_{ox} is equal to the dielectric constant divided by the thickness of the dielectric. Due to the high-K dielectric, we can use thicker dielectrics without facing severe gate leakage, as presented in Fig. 2.9(a). Figure 2.9(b) presents the relationship between gate voltage and source-to-gate leakage. On the left is the PMOS transistor, and on the right is the NMOS device. The black line is the 65-nm Intel process technology displayed on the 2004 IEDM, and the blue line is the 45-nm Intel process technology displayed on the 2007 IEDM [9]. It uses high-K dielectrics and metal gates. We can see that the NMOS gate leakage is reduced by about 25 times. For PMOS, the gate leakage current is increased by 1000 times, which is three orders of magnitude. Therefore, for PMOS, the gate leakage is actually negligible. For those who want to use a large MOS capacitor as a loop-filter capacitor in a PLL, although the unit capacitance is relatively large, it is important to pay attention to the problem of gate leakage. Otherwise, thick oxide devices without gate leakage can be used, some compensation scheme that can compensate for leakage can be applied.

2.3.5   Leakage Mechanism's Dependence on PVT Changes and Possible Methods to Reduce Leakage

2.3.5.1   Voltage Dependence

From the left illustration in Fig. 2.10, it can be seen that the subthreshold leakage and gate leakage are functions that are very dependent on the operating voltage. Therefore, one of the techniques to reduce leakage is to operate the entire circuit or part of the circuit (circuit island) at a lower voltage. For example, if the nominal voltage is

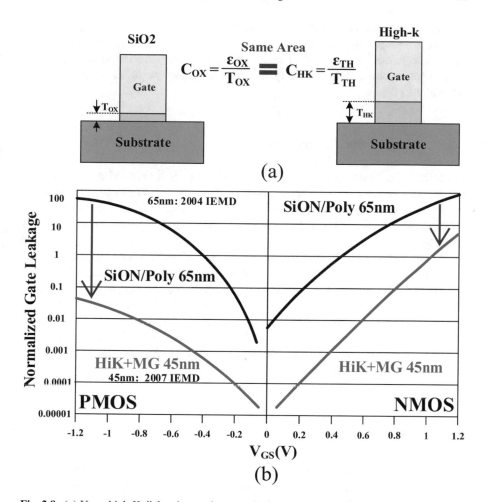

Fig. 2.9 (a) Use a high-K dielectric to reduce gate leakage current. (b) Result of actually reduced gate leakage current

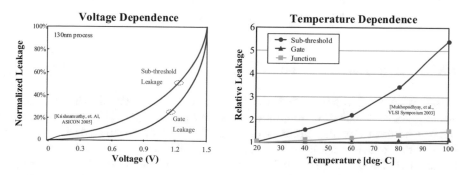

Fig. 2.10 Dependence of subthreshold leakage and gate leakage on voltage and temperature

1.5 V, we can operate the circuit or part of the circuit at 1.2 V, the gate leakage current is reduced to 1/3 of its nominal value, and the amount of subthreshold leakage is halved. In addition, leakage can be significantly reduced. It can also reduce active power, which is why the voltage of noncritical circuit parts should be reduced [10].

2.3.5.2   Temperature Dependence

From the right illustration in Fig. 2.10, it can be seen that the subthreshold leakage dependence is also a very strong function of temperature. When our chip normally works between room temperature and 100 °C, we find that the leakage has increased by five times. Another way to reduce leakage is to lower the operating temperature of the chip.

We can either invest in cooling system research to obtain better thermal interface materials or use multi-core circuits to distribute power consumption and lower the operating temperature of the chip to reduce leakage. The cooling system also helps improve the reliability of the chip. It can be seen from this figure that if the operating temperature is lowered from 100 °C to 80 °C, the leakage can be almost reduced by half, which is a great benefit. Another leakage component, namely, junction leakage, is a weak function of temperature. We find that the gate leakage is almost flat and has little dependence on temperature [11].

2.3.6   Comprehensive Technical Library and Model Card

Generally speaking, in the technology library or model card, there are multiple mixed oxide devices that can work under different power-supply voltages, multiple threshold voltages, or different thicknesses. This is almost an established fact. According to the different requirements of the core or peripheral circuits or I/O circuits to be used, designers can have a high degree of freedom to achieve high-performance (HP) or low-power (LP) design goals.

2.3.6.1   Library: Mixed Oxide

Mixed oxide, the thickness of the oxide is related to the characteristics of the device. All key module circuits will adopt the thin oxide layer design of HP transistors, and other noncritical module circuits will adopt the thick oxide layer design of LP transistors. The advantage of this technology is that it not only has HP transistors but also LP transistors. We can reduce the leakage by 50–200 times while creating a high-speed design. LP and HP transistors can also use different power-supply voltages but need to add level shifting in the interface.

2.3.6.2   Library: Mixed V_T

Hybrid V_T can be a trade-off between the two different design goals of controlling speed and reducing leakage. HV_T denotes lower leakage and slower speed, and LV_T denotes faster speed and more leakage. One of the advantages of using hybrid V_T is that it does not cause area loss and is compatible with standard processes; thus, it is easy to implement at the design level or RTL level.

2.3.6.3   Long Channel Library

Long channel libraries are another way to reduce leakage. The idea is to increase the length L of the transistor. Therefore, there is a loss of area. An example of a transistor whose L value increases by about 10% is to do so without moving the contacts. The speed of this long-channel transistor is about 10% slower, and the leakage is reduced by about three times [12]. The original design procedure is only used long-channel devices, and only nominal components are used in the parts that need to increase the speed.

2.3.7   Bias Technology

The commonly used bias technique is body bias, because the V_T of the transistor is equal to the intrinsic V_T (V_T0) of the transistor, plus a function based on the source body bias V_{SB}.

$$V_T = V_{T0} + \gamma \left[(2\phi_f + V_{SB})^{1/2} - 2\phi_f^{1/2} \right] \tag{2.16}$$

Therefore, we can adjust the threshold voltage to suit our design and application requirements when we adjust the body voltage. Only a low current is required to drive the body bias voltage. For the PMOS, the body bias voltage needs to be higher than V_{DD}. Because the PMOS is placed in well, it is easy to implement. However, for the NMOS, the body bias voltage needs to be lower than the ground voltage, because the body of the NMOS is connected to the substrate, making it difficult to implement. Usually, we only apply bias technology to PMOS.

Another bias technique is source bias, which means that we modulate the source voltage instead of the body voltage. But it needs a high current drive, so it is rarely used.

For the NMOS, when V_{BS} is negative, we call it reverse bias. At this time, NMOS's V_T will increase, but there is a limitation: excessively negative body voltage may cause the breakdown of the gate oxide layer, and the increase in V_{GB} will cause the gate leakage current to increase. The benefit of reverse bias is the

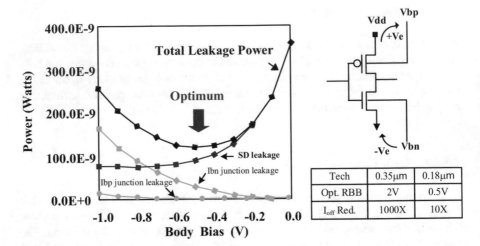

Fig. 2.11 Scalability of reverse body bias

reduction in PN junction leakage and subthreshold leakage. The required cost is a voltage controller and some switches.

Contrarily, when V_{BS} is positive, we call it forward bias. At this time, NMOS's V_T will decrease, but there is a limitation: due to the forward bias of the junction diode, there is a risk of latch-up, so we cannot apply a larger V_{BS}. The advantage of forwarding bias is that V_T is reduced, enhancing the performance of the transistor. The result is that we now have more leakage, and voltage controllers and switches are needed.

Reverse body bias (RBB) has some advantages in minimizing leakage in standby mode [13]. shows the effect of process expansion and reverse body bias. In Fig. 2.11, we treat the leakage power as a function of the applied reverse body bias. As we apply larger body biases, the junction leakage increases, and the subthreshold leakage or source-drain leakage decreases. The best value can be obtained in the total leakage curve. For the 0.18-µm technology, the RBB value is approximately −0.5 V. However, if older technologies such as 0.35 µm are used, the optimal voltage point is higher, and the leakage reduction of the body bias voltage will be significantly better. The trend observed here is that as we scale to smaller and smaller geometric shapes, the body bias effect gradually weakens. This is the reason why body bias is not widely accepted in the industry.

Another way to reduce leakage is to use transistor stacks. Leakage in stacks with multiple turn-off transistors is significantly reduced. It has a similar effect to source biasing. For example, in Fig. 2.12, two NMOS devices are connected in series as a control logic or to create an inverter. If the device is turned off, a residual voltage will be observed at the source of M1. Compared with the gate voltage of M1, the remaining voltage on the source of M1 produces a negative gate-source voltage.

Fig. 2.12 Stacked CMOS transistors

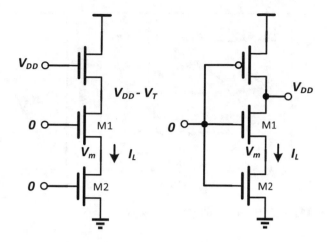

$$V_{\text{GS\_M1}} = -V_{\text{m}} \tag{2.17}$$

$$I_{\text{L}} = I_0 \exp\left(\frac{-V_{\text{m}} - V_{\text{t}}}{nU_{\text{T}}}\right) = I_{\text{L0}} \exp\left(\frac{-V_{\text{m}}}{nU_{\text{T}}}\right) \tag{2.18}$$

The I_{L} leakage is determined by M1, and the negative gate-source voltage has actually a great influence on the stack leakage, which exponentially increases. If we stack transistors, leakage can be reduced by an order of magnitude. Of course, it will also affect speed performance, but it is a very effective way to mitigate leaks.

2.3.8 Voltage Scaling to Obtain Different Energy Performance

As aforementioned, reducing the voltage can also reduce the dynamic power, but the static power, that is, the leakage current, will increase, resulting in the best (or lowest) operating point on the total energy curve, as presented in Fig. 2.13 [14]. Assume that this curve is the result of the energy required by the circuit, such as a ring oscillator as simple as an inverting chain or a NAND chain. On the left-side graph, we plotted the energy required for each operation on the y-axis and the power-supply voltage on the x-axis. When we reduce the power-supply voltage, the kinetic energy follows the CV^2 relationship and continues to decrease. The energy will decrease as expected, but then, it will reach the optimal point—the corner point, which is called the optimal voltage V_{opt}. If the voltage continues to drop, the static energy increases because the leakage power is multiplied by a longer time to perform the task. This means that the total energy will be dominated by static energy and continue to increase.

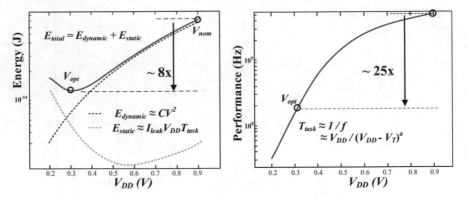

Fig. 2.13 Voltage scaling for energy gain

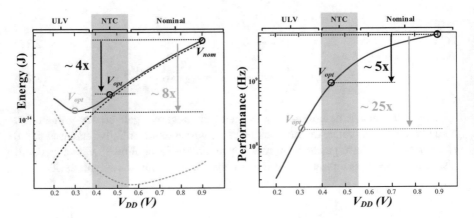

Fig. 2.14 Trade energy for performance

The graph on the right shows the relationship between performance and power-supply voltage. This is a logarithmic y-axis, and performance exponentially drops quickly. Because if it is not for the static energy (leakage curve) to explode upwards, we will continue to follow the quadratic curve, and we will gain energy at a lower voltage. It can be deduced that the optimal point of energy is limited by leakage. In this example, if we look at the numbers, the energy from the V_{nom} point to V_{opt} is reduced by about 6 times, but the performance is reduced by 30 times. It shows the minimum value of V_{opt}, but on the performance vs. voltage curve, it can be observed that the slope is much larger, and it is all in logarithmic space. This means that under this supply voltage, performance is much more sensitive to energy.

Therefore, to make the circuit work at a more excellent and more stable performance, as presented in Fig. 2.14 [1], it is recommended to operate at a slightly higher voltage and discard the energy gain at a lower voltage, so considerable performance will be obtained. In this example, if the energy gain of eight times is returned to two times, the performance loss will be reduced by more than half. A better balance can be achieved in terms of performance loss and energy improvement. This technique of weighing voltage and performance is also called near-threshold design or near-threshold calculation (NTC).

2.4 Process Technology Evolution and Analog Circuit Design

Nanotechnology brings benefits to digital circuits in terms of area and speed, as well as to analog circuits. We can implement more analog function blocks (such as PLL and RF) and integrate them on the same SoC. The total power consumption at the system level is reduced, and then, higher and higher integration levels can be achieved. Due to device expansion, performance continues to improve, and costs continue to drop. Essentially, if the transistor has a shorter gate length, larger G_m, and smaller size, it will end up with more transistors and denser interconnections. In some cases, when we are not limited by noise, we will get the same scaling as digital circuits. But if we are limited by the SNR and have a switched capacitor system, we may actually encounter a certain minimum capacitor size. Therefore, no matter how large the process is scaled, a large analog area may eventually be maintained.

2.4.1 Low-Voltage and Low-Power Analog Circuit Design

In terms of power-supply voltage, according to the ITRS roadmap for the future development direction, the power-supply voltage drops faster than the threshold voltage, which means that the difference between the threshold voltage and the power-supply voltage or the signal swing will also decrease over time. Therefore, as analog designers, we have to worry about this reduction in voltage swing. To make matters worse, the analog circuit must use the same supply voltage as the digital circuit under certain conditions.

Unlike the speed pursued by digital circuits, analog circuits are designed to process analog signals. What we care about is the signal-to-noise ratio (SNR). To obtain the required SNR, the analog signal processing circuit consumes power to convert the signal energy to stay above the basic thermal noise. Therefore, the analog circuit design has a minimum power consumption limit.

For analog circuits, we are actually consuming power to keep the signal above the basic thermal noise ($4kTR_{ON}$) limit. Assume that the OTA efficiency we see in Fig. 2.15 is 100% and completely linear. This means that all the current I_{DD} provided by the supply voltage is used to charge the capacitor C at the output.

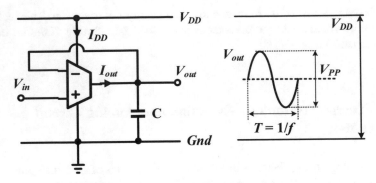

Fig. 2.15 Lower limit of power consumption

$$\overline{I_{\text{out}}} = I_{\text{DD}} = fCV_{\text{pp}} \tag{2.19}$$

$$P = \overline{I_{\text{out}}} \cdot V_{\text{DD}} = fCV_{\text{pp}}V_{\text{DD}} = \frac{V_{\text{DD}}}{V_{\text{pp}}} fCV_{\text{pp}}^2 \tag{2.20}$$

Therefore, the average output power consumption of the OTA is given by Eq. (2.20), which is proportional to the square of the frequency, capacitance, and peak-to-peak voltage V_{pp} [15].

Most people know that the thermal noise power spectral density of the resistor is $4kTR_{ON}$. Therefore, the thermal noise power spectral density of the transistor in the OTA is equal to

$$S_{\text{N}} = 4kT \cdot g_{\text{m}} \cdot \frac{2}{3} = 4kT \cdot g_{\text{m}} \cdot \gamma \tag{2.21}$$

where γ denotes an unnecessary noise factor. To simplify the analysis, we set it to 1.

For switching capacitor systems, the noise voltage across the capacitor is proportional to g_{m}. In this way, when we integrate the noise bandwidth according to the g_{m}/C ratio, g_{m} will cancel out, and the rest is just a simple expression. This is the RMS noise voltage, kT/C:

$$v_{\text{n}}^2 = \frac{kT}{C} \tag{2.22}$$

where C denotes the load capacitance.

For submicron transistors with shorter or deeper gate lengths, they tend to become noisier. For example, for flicker noise (1/f noise), the square of its noise voltage is equal to $K_{\text{F}}/(C_{\text{OX}}^2 WLf)$. The deeper the submicron, the greater the K_{F}. In high-speed circuits or broadband circuits, the proportion of low-frequency noise to the total noise is not significant. However, if we design a circuit that works in the audio range, we should consider this issue. Here, we first study the noise generated by

OTA, considering only thermal noise and ignoring flicker noise. Therefore, the SNR can be simply given by the RMS value of the signal V_{pp} equation:

$$P_s = \frac{V_{FS}^2}{8} = \frac{V_{pp}^2}{8} \qquad (2.23)$$

Then, we can express the peak-to-peak voltage required to reach a specific SNR at a given capacitance C and a given temperature. This enables us to obtain expressions for power, SNR, bandwidth, and voltage, which is a real compromise.

$$\text{SNR} = \frac{V_{pp}^2/8}{kT/C} = \frac{V_{pp}^2 C}{8kT} \rightarrow V_{pp}^2 = 8\frac{kT}{C} \cdot \text{SNR} \qquad (2.24)$$

$$P = 8kTf \cdot \text{SNR} \frac{V_{DD}}{V_{pp}} \qquad (2.25)$$

This indicates that at a given temperature, the average power is proportional to the frequency or bandwidth. It also depends linearly on the SNR. If $V_{pp} = V_{DD}$, we can get the lowest power consumption of the analog circuit.

$$P_{min} = 8kTf \cdot \text{SNR, if } V_{pp} = V_{DD} \qquad (2.26)$$

Therefore, we can immediately see that if we want to minimize power consumption at a given bandwidth, temperature, and SNR, the rail-to-rail operation is required to fill the complete supply voltage.

All performance indicators of analog circuits are actually declining. Why do we still need low-voltage analog circuits? This is actually to reduce the dynamic power consumption ($f \cdot C \cdot V_{DD}^2$) in the digital circuit. The lower power-supply voltage also limits the electric field in the device to avoid any strong short-channel effect process. Therefore, when we want to perform complete system integration, we take the digital circuit part as the dominant one. Although we still have some analog circuits or RF circuits, we may want to integrate the same system on the chip, but we do not want to generate too many different power-supply voltages. Therefore, we may face low supply voltage. Another situation is, for example, we are using an energy harvesting source. Generally, these energy sources provide very small voltages and force us to use power-supply voltages that are actually very small.

Let us now study the impact of power-supply voltage reduction on power consumption. Figure 2.16 presents a simple NMOS transconductance gain stage. It has NMOS transconductance with PMOS current source. We assume that they all have the same saturation voltage V_{DSsat}, and we need to process the signal peak-to-peak amplitude V_{pp} under a certain V_{DD} voltage.

First, let us assume that V_{DSsat} is incompressible, so we keep them unchanged. V_{pp} can be obtained by the simple equation listed below

Fig. 2.16 Influence of power-supply voltage drop on power consumption

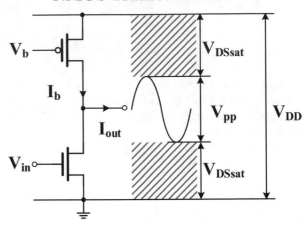

$$V_{DD} = V_{pp} + 2V_{DSsat} \rightarrow V_{pp} = V_{DD} - 2V_{DSsat} \qquad (2.27)$$

Therefore, V_{pp} is the difference between V_{DD} and $2V_{DSsat}$. Then, we can substitute it into Eq. (2.25) to obtain the power consumption of the analog circuit.

$$P_{analog} = 8 \frac{V_{DD}}{V_{DD} - 2V_{DSsat}} kT \cdot f \cdot SNR \qquad (2.28)$$

Now, we discuss the two variable parameters in Eq. (2.28): power-supply voltage and SNR.

Power-supply voltage: Low power-supply voltage will definitely bring difficulties to analog design because everything is reduced. The available range will become smaller and smaller until it eventually disappears. The lower the power supply, the fewer cascode that can be applied, and the more difficult it is to obtain voltage gain. The amplifier specifications have input and output ranges. The input range is not a big problem, and a smaller input range can be used. The reason for this is that the amplifier uses feedback; thus, the inputs are driven together to virtual ground. At the output, the situation is completely different. Due to the reduction in power, the drive capability of the amplifier is limited, as well as the output range.

It can be observed that as we reduce V_{DD} and approach $2V_{DSsat}$, the signal space or V_{pp} gets smaller and smaller. As presented in Eq. (2.28), reducing V_{DD} will result in an increase in analog power.

SNR: In order to maximize the SNR, on the signal side, we must somehow maximize the signal swing within the power limit we have. Therefore, we must modify the topology and operation, which means that we should not waste too much headroom in the circuit. For example, we can remove some cascode transistors to increase the swing. In terms of noise, we must minimize noise contributors and their contributions. For example, in PLL, we prefer passive loop filters to active loop filters, so there will be fewer noise contributors there. If noise reduction technology is used, the SNR is still not sufficient, and finally a higher V_{DD} can only be used to increase the signal amplitude.

If the circuit does not request high SNR, for example, the SNR performance requirement of a 6-bit flash A/D converter, the faster the circuit clock, the more power it consumes. In this case, the converter behaves almost like a digital circuit. However, when the circuit requests high SNR, the reduction in power may actually result in higher power consumption. If a switched capacitor system is used, the kT/C noise is determined by the size of the capacitor. To maintain the SNR, when the power-supply voltage drops, it is important to reduce noise and increase capacitance. For example, to reduce the noise by a factor of 2, we must increase the capacitor by a factor of 4. In fact, it may consume more current or even more power than a higher supply voltage.

To design a low-supply-voltage analog circuit, we must accept the fact that the peak-to-peak voltage and V_{DSsat} continue to drop. This means that we must shift the operating point from a strong inversion to a moderate inversion and finally perform it in a weak inversion. Traditional methods also need to be replaced by the new G_M/I_D design method to improve the efficiency of the transistor, as described below.

2.4.2   Square-Law and G_M/I_D Design Methods

Square-law design method: When designing an analog circuit, the transistor size needs to be adjusted according to the target specification. The target specifications gain-bandwidth product, minimum area, low-voltage design, minimum power consumption, dynamic range, etc., are very complex and difficult tasks. In deep submicron design, the ideal square-law model of CMOS transistors we learned in school,

$$I_D = \mu C_{OX} \frac{W}{L} (V_{GS} - V_T)^2 (1 + \lambda V_{DS}), V_{DS} \geq V_{GS} - V_T \geq V_T \qquad (2.29)$$

$$g_m = \sqrt{2\mu C_{OX} \frac{W}{L} I_D} \qquad (2.30)$$

$$g_{DS} = \lambda I_D, \lambda \approx L \qquad (2.31)$$

One advantage of square-law equations is that they are easy to derive from basic solid-state physics. Algebras are simple, but they are very useful for gaining insight into the basic CMOS circuit behavior. Therefore, the square-law model is still very useful as a "warm-up tool" for circuit design students. However, it is very difficult to calculate the size of the transistor using this method because it is not easy to obtain device parameters of the transistor model, such as μ, C_{OX}, and λ. Furthermore, the square-law MOS model suffers from several limitations, especially when it involves short-channel transistors and medium inversion, which is completely invalid in weak inversion (subthreshold operation).

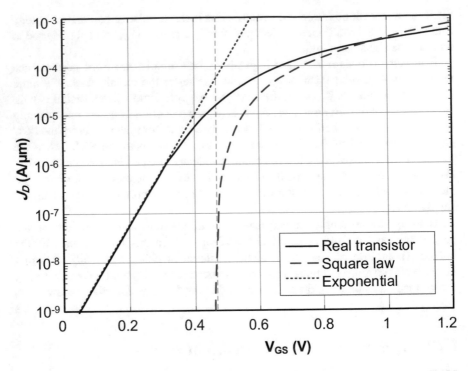

Fig. 2.17 Current density and VGS of the smallest length n-channel device in 65-nm CMOS technology

In Fig. 2.17, the current density map of a real 65-nm transistor is clearly visible (where the vertical dashed line corresponds to the threshold voltage of the device). It shows the approximate value of the exponent and the law of squares. The exponent provides a fairly good fit at a very low V_{GS} (weak inversion), and the quadratic curve approximation begins to make sense at a few hundred millivolts above the device threshold voltage (vertical dashed line). Ideally, the transition from weak to strong inversion should be smooth and continuous, but it turns out that it is not easy to find the physical relationship between the exponent and the law of squares. This is because the square law of hand calculation described in textbooks is usually inconsistent with the classical simulation process. Modern circuit simulation relies on complex device models, such as BSIM6, which have been carefully designed to achieve the "real" device characteristics in Fig. 2.17. Therefore, there is a clear disconnect between manual analysis and simulation results. Moreover, designers usually avoid manual calculations and instead adopt an iterative and time-consuming SPICE-based "adjusted" design style. Using the G_M/I_D method can reduce this dilemma.

2.4.2.1 G_M/I_D Design Method

G_M/I_D is based on the classic manual analysis method and uses a lookup table generated by SPICE to eliminate the gap between manual analysis and complex transistor behavior. These tables contain multidimensional sweeps of transistor equivalent to small-signal parameters (G_M, G_{DS}, etc.) across the MOSFET terminal voltage. Since the lookup table data can be used to closely capture the behavior of the SPICE model, close consistency between the required specifications and simulation performance can be achieved.

It is important to briefly describe the method of G_M/I_D and f_T parameter simulation design to determine the size of the CMOS analog circuit to meet specifications such as gain bandwidth while optimizing features such as low power consumption and small area. Two parameters that seem very important to analog designers are (1) G_M/I_D and (2) $f_T = G_M/(2\pi C_{GG})$. The former represents the amount of current to be used by each transconductance, whereas the latter represents how much total gate capacitance C_{GG} must be driven at the control node for each required transconductance. Our objective is to enable the MOS transistor to obtain a large G_M without a large amount of current I_D and C_{GG}. The use of this design method can help us achieve our objective. The value of G_M/I_D and f_T depends on the working area of the MOS tube. As presented in Fig. 2.18(a), the G_M/I_D parameter has the maximum value in the weak inversion region and decreases as the operating point moves to the strong inversion region. In Fig. 2.18(b), the f_T parameter has a minimum value in the weak inversion region and increases as the operating point moves to the strong inversion region. In addition, it has been observed that the G_M/I_D parameter is not very sensitive to technology scaling. On the other hand, the value of f_T significantly increases with the scaling of technology. The analog design engineer is more concerned about the product of G_M/I_D and f_T, as presented in Fig. 2.19 [16]. According to the design goal, the most suitable operation region and overdrive voltage, namely ($V_{GS} - V_T$), should be chosen to achieve the bandwidth goal while operating with the maximum possible G_M/I_D (lowest power). The overdrive voltage

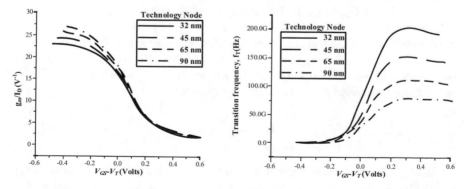

Fig. 2.18 Simulation results of g_m/I_D and f_T varying with the operating regions and technology nodes [16]

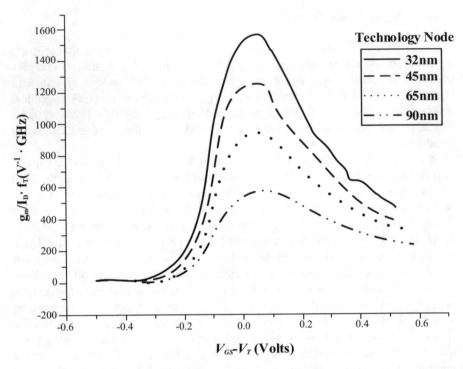

Fig. 2.19 Simulation results of the product of g_m/I_D and f_T varying with the operating regions and technology nodes [16]

shows the "sweet spot" around 100 mV, which corresponds to the common bias conditions in many mid-to-high-speed designs. On the other hand, in low-voltage or low-power analog circuits, the MOS transistor can be biased in the weak inversion region, and the use of high G_M/I_D greatly helps in reducing the power consumption of applications that do not require extremely high bandwidth.

The choice of G_M/I_D as the low-voltage analog circuit design is based on the following three reasons. First, this parameter is closely related to the simulation performance. Second, it represents the operation region of the MOS device. Third, it provides tools for calculating the size of transistors. Since the ratio of G_M/I_D does not depend on the gate width, the drain current up to any specified gain-bandwidth product can be derived from the following equation:

$$I_D = \frac{G_m}{(G_m/I_D)^*} \qquad (2.32)$$

The numerator is determined by the design specification $gm = \omega_T \cdot C$. The $(g_m/I_D)^*$ ratio of the transistor in the denominator comes from a similar device, and its gate width W^* and gate length L^* are known.

Knowing the drain currents, widths follow from the proportionality:

$$W = \frac{I_D}{\left(I_{D/W}\right)^*} = \frac{I_D}{\left(J_D\right)^*} \tag{2.33}$$

Equations (2.32) and (2.33) form a set of parameter equations to determine the drain current and gate width to achieve the gain-bandwidth product fixed by g_m. The key to the scale calculation method is the denominator of Eq. (2.32). Because it functions as a parameter, it can perform transistor scan in all operating modes. It is actually the slope of the drain current characteristic relative to the gate voltage plotted on the semi-logarithmic axis, where

$$\left(g_m/I_D\right)^* = \frac{1}{I_D^*}\frac{dI_D^*}{dV_G} = \frac{d}{dV_G}\log\left(I_D^*\right) \tag{2.34}$$

The main advantage of the G_M/I_D method compared with other methods is that it uses a set of lookup tables as the main design tool, which itself is constructed based on the SPICE simulation. Therefore, the predicted results obtained after the circuit sizing process seem to be very close to the actual SPICE results. However, the construction of the lookup table is much easier and less time-consuming. Therefore, the G_M/I_D method is more and more important for nanoscale analog circuit design. In addition, the G_M/I_D method allows designers to flexibly operate transistors in any operation region.

The general process of determining the size of a transistor is summarized as follows:

Determine G_M (according to the design specification $g_m = \omega_T{\cdot}C_L$).
Choose L (use the smallest available channel length L):
Short passage $\rightarrow$ high speed, small area
Long channel $\rightarrow$ high internal gain, improved matching
Choose G_M/I_D:
Large $G_M/I_D \rightarrow$ low power consumption, large signal swing (low V_{DSsat})
Small $G_M/I_D \rightarrow$ high speed, small area
Determine I_D (from g_m and g_m/I_D).
Determine W (from I_D/W).

Practice example: Size the circuit of Fig. 2.20(a) so that $f_T = 1$ GHz when $C_L = 1$ pF. Assume that $L = 60$ nm, $g_m/I_D = 15$ S/A (moderate inversion), $V_{DS} = 0.6$ V, and $V_{SB} = 0$ V.

Solution:
Following the aforementioned procedure presented in Fig. 2.20(a), we start by computing the transconductance:

$$g_m = 2\pi f_T C_L = 6.26 \text{ mS}$$

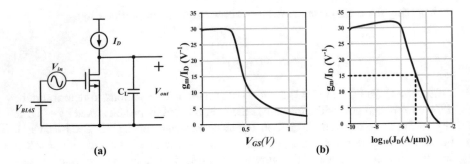

Fig. 2.20 (a) Circuit schematic of an intrinsic gain stage. (b) g_m/I_D curves of $L = 60$ nm [17]

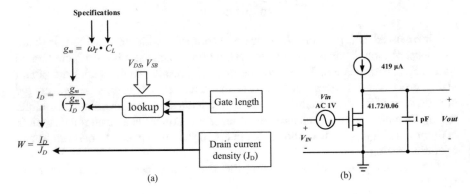

Fig. 2.21 Schematic of the solution for the practice example

Since g_m/I_D is given and equal to 15 V^{-1}, we can find I_D using Eq. (2.32):

$$I_D = \frac{g_m}{(g_m/I_D)^*} = 6.28 \text{ mS}/15 = 419 \text{ uA}$$

To find the width W, we divide I_D by the drain current density J_D and find the drain current density at $g_m/I_D = 15$ V^{-1}, as presented in Fig. 2.20(b).

We find that $\log_{10}(J_D) = -4.998$, and J_D is equal to 10.05 µA/µm. Therefore, calculate I_D/J_D to get $W = 41.72$ µm, so we have completed the size adjustment process. We see from this process that once we fix g_m/I_D, the drain current I_D is known. According to the provided gate length, W can be determined by looking up the table, and the circuit implementation is shown in Fig. 2.21(b).

2.5 Implementation and Precautions of Nanometer Analog Circuit Design

2.5.1 Process Variability Problem

Although the feature size is decreasing, the critical size control in the process is not expanding at the same speed. For example, starting from the 180-nm node, the wavelength of ultraviolet radiation used for photolithography has been lower than the drawn gate length. In addition, random doping fluctuations during the manufacturing process can cause a significant increase in the threshold voltage variation. As the power-supply voltage decreases, V_{GS}–V_T must be kept small, rendering the current source matching poor. And the leakage current also increases with considerable variability. The end result is a huge difference in characteristics from wafer to wafer. Due to photolithography, manufacturing plants has become increasingly difficult. To improve the yield, more and more complex layout rules have been added. More DRC rules should be included for smaller geometries. There are also some rules, such as proportional space and width rules as well as density and filling rules.

For accurate matching, dummy gates and devices need to be added. These dummy gates and devices also increase the parasitic effects of the circuit and the complexity of the analysis. To match the layout (such as common-centroid layout type), wiring and additional parasitic capacitance may be added, and the characteristics of the circuit may also be changed, which may cause greater mismatch. To minimize variability, we must use some kind of adaptive biasing technique, or we can use some kind of calibration method to avoid circuit variability.

2.5.2 Signal Integrity Issues

With the reduction of the minimum feature size of devices, the advancement of underlying semiconductor technology, and the demand for higher clock frequencies and higher analog accuracy, signal integrity issues have become more and more important in the design of mixed-signal integrated circuits. Signal integrity consists of three parts: signal line integrity (SI) is the interaction between signal line and signal reflection and attenuation caused by impedance mismatch during transmission. Power integrity (PI) is a fast digital transient switching noise that couples to adjacent circuits through nonideal power/ground planes, causing resonances and discontinuities in the power/ground return path. Electromagnetic interference (EMI) is a high-frequency common-mode or differential-mode radiation that can cause interference to remote circuits.

The miniaturization of the process enables every part of the signal integrity to be involved with each other, because the distance between the lines and the blocks is closer, and the chance of coupling from one end of the circuit to the other is greater.

Therefore, cross talk occurs. In addition, in the low-voltage design of the nano-CMOS process, since the signal is already very small, special attention should be paid to every noise, especially the power-supply and substrate noise. In a submicron system or large-scale mixed signal, the power supply and substrate are connected to each circuit block, so the noise generated in any block will be coupled to other circuits or blocks. And as the clock frequency increases, the substrate noise will worsen. This is true for both digital and analog circuits. The reason is that for higher or faster rise times, the transient will not completely disappear. As more transistors work at the same time, the power supply needs to provide higher current and higher d_i/d_t. If the connection between the power supply and the ground is insufficient, or there is any type of inductance between the power supply and the ground, there will be a voltage drop and limit performance.

The methods to alleviate this problem are as follows:

1. The key module adopts an on-chip voltage regulator. The requirement for the regulator is good power-supply rejection ratio (PSRR). To reduce power-supply noise, it is not sufficient to allocate separate V_{DD} and V_{SS} pins for key modules. A large number of power-supply pins are placed close to the edge of the chip (10 power supplies are a basic requirement for general System-on-chip (SoC)). It is important to use a very wide power bus and keep the resistance around 1 Ω (>20 μm if possible). The wide power bus helps reduce cross talk, and has very low impedance and very good decoupling. In fact, many problems will be easily solved, and even the substrate noise will be very limited. If possible, multiple welding wires should be used. Low-inductance bonding (<2-nH target) can significantly improve performance. As performance improves, the cross talk is actually limited.
2. Large-area distributed on-chip decoupling capacitors (tens of pF) are used. On-chip decoupling essentially provides us with a charge buffer, which can accomplish two things. If we have a current spike, it can be removed by the charging buffer to stabilize the voltage. Another thing is that it limits the noise from other blocks, similar to the RC filter function. MOS capacitors can be used as decoupling capacitors because of their high density. However, due to gate leakage issues, if the leakage exceeds your tolerance, the use of core MOS devices should be avoided.
3. It is recommended to adopt a fully differential structure, and the CMRR can be increased by more than 10 dB. It should not be expected that high-precision single-ended systems can be made using large chips and work well.

2.5.3 Implementation Considerations

2.5.3.1 Analog Switch

To realize CMOS analog switch, NMOS and PMOS can be used in parallel. As long as the sum of V_T is less than the value of V_{DD}, any voltage can be switched within the

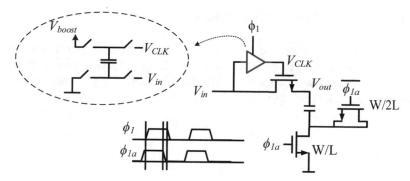

Fig. 2.22 Boosted sampling switch and dummy switch

entire V_{DD} range. In addition to requiring the MOS itself to have a constant on-resistance, it should also be ensured that V_T is low enough to switch fast enough. However, the disadvantage of lower V_T is that when we turn switches off, they may not be completely off, like switches with higher V_T, but instead start to leak.

In many low-voltage circuits, when using a sampling switch, it is important to avoid injecting charge into the capacitor. It affects the common-mode range, causing gain errors and distortion. It is desirable to minimize the injected charge. A common method is to place a virtual switch. Take the bottom-plate sampling technique as an example, as presented in Fig. 2.22. If we use a switch to charge the capacitor, the bottom-plate switch ϕ_{1a} should always be turned on. Moreover, a virtual switch should be placed, and the opposite phase of ϕ_{1a} should be used to balance the charge that we should have. Compared with traditional switches, the size of virtual switches can reach W/2. In addition, in low-voltage circuits, if the V_T sum is less than V_{DD}, a boost switch must be used to minimize input distortion. Furthermore, NMOS should be used, and a switch voltage proportional to the input voltage should be used to keep the impedance of the switch as constant as possible. The clock booster can be implemented using a simple switched capacitor circuit. The boost capacitor should be charged to a given voltage and then superimposed on the input voltage when ready to sample.

More and more metal capacitors are used in deep submicron processes. People who used metal capacitors in the past usually stacked them vertically. But today, photolithography technology makes the horizontal distance between metal lines actually smaller than the vertical distance between metal plates and metal lines. By using horizontal line (finger) structure capacitors that are very close to each other, capacitors (up to 1 fF/μm^2) with very high density and excellent matching (10–12-bit ADCs do not require additional processing steps to trim) can be obtained.

Resistors are expected to have high resistivity so that they can become smaller. They also need to have good temperature stability. Many processes do not provide precise resistors because they are difficult to control, and the absolute variation may be $\pm 20\%$. If there is no need to use resistors, it is best to avoid them, or resistors should only be used for voltage dividers.

2.5.3.2 Only Use Current When Needed

All unnecessary functions must be disabled. We need to turn off unnecessary functions in the circuit to prevent the system from supplying power to a large number of nonworking components. For LP biomedical applications, this is an absolutely necessary principle.

References

1. Pinckney, N., Blaauw, D., & Sylvester, D. (2015). Low-power near-threshold design: Techniques to improve energy efficiency energy-efficient near-threshold design has been proposed to increase energy efficiency across a wid. *IEEE Solid-State Circuits Magazine, 7*(2), 49–57.
2. Dennard, R. H., et al. (1974). Design of ion-implanted MOSFET's with very small physical dimensions. *IEEE Journal of Solid-State Circuits, 9*(5), 256–268.
3. Akbar, M. S. (2005). *Process development, characterization, transient relaxation, and reliability study of HfO_2 and $HfSi(x)$ $O(y)$ gate oxide for 45nm technology and beyond.* Doctoral dissertation.
4. Takahashi, M., et al. (1998). A 60-mW MPEG4 video codec using clustered voltage scaling with variable supply-voltage scheme. *IEEE Journal of Solid-State Circuits, 33*(11), 1772–1780.
5. Chandrakasan, A. P., Sheng, S., & Brodersen, R. W. (1992). Low-power CMOS digital design. *IEICE Transactions on Electronics, 75*(4), 371–382.
6. Zhang, K., et al. (2005). SRAM design on 65-nm CMOS technology with dynamic sleep transistor for leakage reduction. *IEEE Journal of Solid-State Circuits, 40*(4), 895–901.
7. Rusu, S., et al. (2005, June 30). *Cache leakage shut-off mechanism.* US Pat US20070005999A1.
8. Henzler, S., et al. (2005). Sleep transistor circuits for fine-grained power switch-off with short power-down times. In *ISSCC. 2005 IEEE International Digest of Technical Papers. Solid-State Circuits Conference*, pp. 302–600.
9. Mistry, K., et al. (2007). A 45nm Logic Technology with High-k+Metal Gate Transistors, Strained Silicon, 9 Cu Interconnect Layers, 193nm Dry Patterning, and 100% Pb-free Packaging. In *IEDM 2007*, pp. 247–250.
10. Krishnamurthy, R., et al. (2005). High-performance and low-voltage challenges for sub-45nm microprocessor circuits. In *International conference on ASIC (ASICON), 2005*, pp. 283–286.
11. Mukhopadhyay, S., & Roy, K. (2003). Accurate modeling of transistor stacks to effectively reduce total standby leakage in nano-scale CMOS circuits. In *VLSI Symposium 2003*, pp. 53–56.
12. Rusu, S., et al. (2006). A dual-core multi-threaded Xeon processor with 16MB L3 cache. In *2006 IEEE International Solid State Circuits Conference-Digest of Technical Papers*, pp. 315–324.
13. Keshavarzi, A., et al. (2001). Effectiveness of reverse body bias for leakage control in scaled dual Vt CMOS ICs. In *Proceedings of the 2001 international symposium on Low power electronics and design, August 2001*, pp. 207–212.
14. Pinckney, N., et al. (2012). Assessing the performance limits of parallelized near-threshold computing. In *Proceedings of the 49th Annual Design Automation Conference*, pp. 1147–1152.
15. Vittoz, E. A., & Tsividis, Y. P. (2002). *Trade-offs in analog circuit design.* Springer.
16. Pandit, S., Mandal, C., & Patra, A. (2018). *Nano-scale CMOS analog circuits: Models and CAD techniques for high-level design.* CRC Press.
17. Jespers, P. G. A., & Murmann, B. (2017). *Systematic design of analog CMOS circuits.* Cambridge University Press.

Chapter 3
Introduction of Frequency Synthesizer

This chapter introduces the basic principles of the integer-N frequency synthesizer and its linear model. Then, some key trade-offs when designing a synthesizer are discussed, including the trade-offs between stability, phase noise, and reference spurs. Next, each building block is introduced in the frequency synthesizer, including the phase detector, charge-pump, loop filter, oscillator, and frequency divider. Next, we briefly introduce non-integer frequency synthesizers, as well as direct synthesizers and all-digital phase-locked loops (ADPLLs). In the ADPLL, it has a one-to-one correspondence with the traditional analog PLL. Because digital filters are well discussed in some textbooks, we mainly focus on the other two important units, namely, digitally controlled oscillation and time-delay conversion. Finally, the design example of the frequency synthesizer is briefly introduced.

3.1 Introduction of Frequency Synthesizer

A frequency synthesizer mainly has two major applications. One is a clock generator used to provide a stable clock for most digital systems. The other is a local oscillator in the radio transceiver; it is used for hopping the frequency of the RF circuit and switching the channel. Every radio is using it, as presented in Fig. 3.1. Every radio has a transmitter/receiver and requires a local oscillator to perform upconversion/downconversion. Furthermore, the frequency of these local oscillators must be tunable and must have sufficient coverage, not only to cover the entire communication channel but also to cover the entire frequency band. There must also be a certain margin to deal with the process voltage and temperature changes. Unlike clock generators, frequency synthesizers must consider spectral purity in addition to timing jitter. For example, the out-of-band phase noise may cause a reciprocal mixing problem for blocking issues, and then, the in-band phase noise may also

C.-C. Hung, S.-H. Wang, *Ultra-Low-Voltage Frequency Synthesizer and Successive-Approximation Analog-to-Digital Converter for Biomedical Applications*, Analog Circuits and Signal Processing, https://doi.org/10.1007/978-3-030-88845-9_3

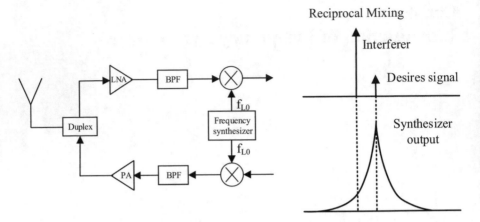

Fig. 3.1 Frequency synthesis in a radio transceiver

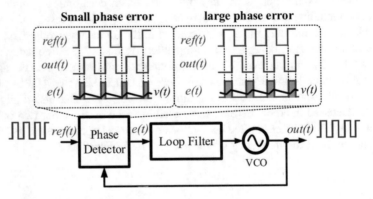

Fig. 3.2 Operation of phase-locked loop (PLL)

cause a high integration phase error, which results in a decrease in bit error rate performance. If the divisor = 1 in the feedback path of the frequency synthesizer, it can also be simplified into a PLL, which can be used in a clock data recovery circuit for clock recovery.

3.1.1 Introduction to Phase-Locked Loop (PLL)

To understand the operation of the frequency synthesizer, we first start with the simplified version of the frequency synthesizer, which is the PLL. The idea of a PLL is that we have some kind of oscillating component, as presented in Fig. 3.2, whose oscillation frequency is controlled by some input voltage. A very precise frequency setting is expected; thus, the oscillator output should be compared with the input reference frequency (such as a crystal oscillator) in phase by creating an output pulse

Fig. 3.3 PLL used as a
high-Q band-pass filter with
auto-tracking feature

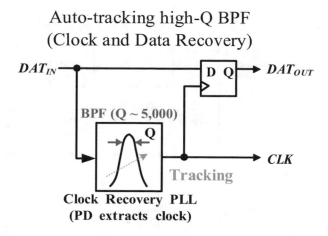

Fig. 3.3 PLL used as a high-Q band-pass filter with auto-tracking feature

from the phase detector, with the pulse width corresponding to the edge-to-edge instantaneous time difference. When the output pulse has a greater pulse width, it means that the phase difference between the two is larger. In order to reduce the pulse width, a loop filter is used to average the pulses sequentially, and the average value will be used as a function of the actual phase difference. Then, the average value is fed directly into the oscillator, and a feedback loop that determines the relationship between the output frequency and whether the actual input frequency is locked. Because the frequency is the derivative of phase, as long as a constant phase error can be produced, a zero-frequency offset can always be obtained through a PLL. This is why phase lock is a technique for achieving frequency lock targets.

Another unique feature of the PLL is that it can provide high-Q band-pass filter devices with high ductility and automatic tracking characteristics, which can be used for clock recovery in clock data recovery circuits to achieve good jitter tolerance, as presented in Fig. 3.3.

3.1.2   Linear Model of PLL

Figure 3.4 presents the linear model of PLL. Although the PLL is a nonlinear feedback system, the behavior of the PLL can be approximated by a linear model within the locked range of interest. Therefore, the open-loop gain and closed-loop transfer function can be easily obtained from the explanation of the basic circuit theory definition.

The type of PLL is the number of integrators in the loop. In other words, in the transfer function gain of the loop, the number of poles at the origin indicates the type of PLL, and the order of the PLL is actually the total number of poles in the transfer function. Thus, the order of the PLL is obviously the number of poles of H(s) or the number of roots of 1 + G(s), so the order is always greater than or equal to this type.

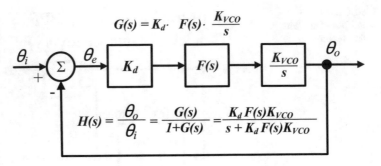

Fig. 3.4 PLL linear model

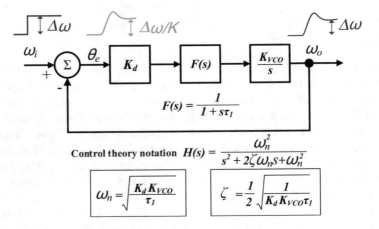

Fig. 3.5 Linearity analysis of Type-1 PLL

For most synthesizer applications, we generally use a linear discriminator, so its transfer function can be expressed in pure gain K_d. The loop filter will have a frequency-dependent transfer function $F(s)$, and the oscillator can be represented by K_{VCO} divided by s (the Laplace variable).

According to the above definition, for Type-1 PLL, we only need one pole, which is given by VCO; the loop filter can be a simple first-order, low-pass filter (LPF), as presented in Fig. 3.5 [1]. The entire closed-loop transfer function is a second-order function. Then, a standard control theory notation can be used to write this function as $\omega_n^2/(s^2 + 2\zeta\omega_n s + \omega_n^2)$, where ω_n denotes the natural frequency of the system, and ζ denotes the damping factor.

Let us now examine the transient behavior of the PLL. In the Type-1 PLL, the integrator is already in the loop, which means that there are no other integrators in the loop. Therefore, when the input signal has frequency hopping, the VCO output must have the same $\Delta\omega$. But the loop filter does not have an integrator, so the PLL cannot store the DC voltage to adjust the VCO output frequency. The only way to provide DC voltage information to the VCO is to create a phase detector with a direct static

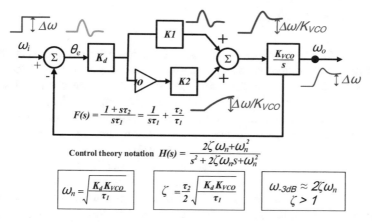

Fig. 3.6 Linearity analysis of Type-2 PLL

phase offset, as presented in Fig. 3.5. Because there is only one VCO in the loop, it is very stable.

However, for Type-2 PLL, in addition to the pole provided by the original oscillator, there is an integrator inside the loop filter to provide another pole, so two paths in Type-2 PLL can be used. One path is the proportional gain path, which is similar to Type-1 PLL. The other path is the integration path, which contains an integrator. Therefore, the integrator can store the DC voltage to adjust the output frequency of the voltage-controlled oscillator. Furthermore, the phase detector does not need to provide a static phase offset to add other DC voltage information. The transfer function of the loop filter is presented in Fig. 3.6. Such a transfer function is a second-order function that can be rewritten as $(2\zeta\omega_n s + \omega_n^2)/(s^2 + 2\zeta\omega_n s + \omega_n^2)$ in the cybernetic notation.

If the damping coefficient is greater than 1, the denominator of the transfer function (called the characteristic equation) can be decomposed into two real roots. In this case, what happens is that the lower one of the two poles is close to zero and will deviate from zero. Therefore, the whole thing can be approximated as a first-order transfer function with a bandwidth of $2\zeta\omega_n$.

A brief summary of the advantages of Type-2 PLL over Type-1 is as follows. With reference to Type-2 PLL, the damping coefficient and natural frequency can be independently selected. Since the position of zero can be selected independently, there is an additional degree of freedom. A more important advantage of Type-2 PLL is that for any phase step or frequency step, the steady-state error is zero. This is why Type-2 PLL is the most popular. In the design of PLLs, third-order, fourth-order, or even higher-order PLLs may be observed. However, the design of these high-order PLLs only adds poles outside the bandwidth of the PLL, which means that the basic characteristics of the high-order PLL and the second-order PLL are almost the same.

3.1.3 Integer-N PLL

If we want to synthesize a clock frequency of N times the input frequency, then we need to divide the output clock by N in the feedback path for so-called indirect frequency multiplication, as the reference frequency will be compared with the divider output frequency. The phase of the two input signals of the phase frequency detector (PFD) must be the same, so the oscillation frequency of the VCO needs to be N times the reference frequency. Since N is any integer, it can be called an Integer-N PLL or an Integer-N frequency synthesizer.

Figure 3.7 presents a typical Integer-N PLL. Basically, it is a Type-2 charge-pump PLL, and all modules can be modeled by corresponding transfer functions. PFD and charge pump are represented by $I_{cp}/2\pi$. In order to make the PLL easier to analyze, it is assumed to be a second-order system. The loop filter equation ignores the high-frequency poles and only has a proportional path and an integral path. The VCO transfer function is K_{VCO} divided by s. The transfer function of the phase-locked loop can be expressed by the damping coefficient and the natural frequency. The figure also shows the expressions of the natural frequency, ω_n, and damping factor. When the damping is greater than 1, the phase-locked loop system can be approximated as a first-order low-pass filter, and the −3 dB frequency can be obtained, as shown in the figure.

Although the model is a linear system, the feedforward path is continuous-time, and the feedback path is discrete-time. The feedforward is composed of resistors, capacitors, and VCO. However, all actions only occur during the small pulse width of either the reference signal or the feedback signal. Between these pulses, no output is generated. This has very important effects on the settable PLL bandwidth. It is a bit like trying to talk to passengers while driving.

Imagine you are driving; your friend is sitting next to you, and you are talking while driving. You want to pay more attention to your friend, but your friend wants

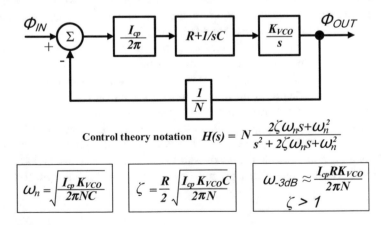

$$\text{Control theory notation} \quad H(s) = N\frac{2\zeta\omega_n s + \omega_n^2}{s^2 + 2\zeta\omega_n s + \omega_n^2}$$

$$\omega_n = \sqrt{\frac{I_{cp}K_{VCO}}{2\pi NC}} \qquad \zeta = \frac{R}{2}\sqrt{\frac{I_{cp}K_{VCO}C}{2\pi N}} \qquad \omega_{-3dB} \approx \frac{I_{cp}RK_{VCO}}{2\pi N} \quad \zeta > 1$$

Fig. 3.7 Typical Integer-N PLL

you to pay more attention to the road because this is related to life safety issues. These are conflicting requirements. So, all you have to do is to look at the road first and then occasionally look at your friend. The frequency of sampling road conditions is equivalent to the bandwidth of the PLL. The larger the bandwidth, the faster the sampling frequency, and vice versa. Now, if the road is very straight, then the feedback system you develop as a car driver when driving on the road does not need to have a very wide bandwidth. In other words, you do not have to constantly adapt to changes in the road, so your bandwidth may be lower. However, if you are on a winding mountain road, you must frequently sample the road. You have to update the feedback system more frequently or widen the feedback system bandwidth. If you spend too much time looking at your friend and you make a mistake, you will be on the road to death. In the PLL, because we only sample at the reference frequency and do nothing between the two samples, the bandwidth cannot be too wide. A too-wide bandwidth forces you to keep issuing corrective commands, but the PLL responds slowly. When the PLL starts to respond, too many correction commands are issued, and correction in the opposite direction is required at this time. Therefore, the upper limit of the bandwidth must be set. Unless you ensure that the bandwidth is less than one-tenth of the reference frequency, the PLL will be unstable.

3.2   Trade-Offs in Synthesizer Design

When designing a synthesizer, three key indicators should be considered: phase noise, reference spurs, and settling time. Since the PLL is a feedback system, these three parameters are highly sensitive to the PLL bandwidth. Therefore, one of the most important tasks for PLL designers is to define the best bandwidth to achieve the best performance by considering three design parameters. But before that, the bandwidth of the PLL must be determined first.

3.2.1   Selection of PLL Bandwidth

If the bandwidth you selected is less than or equal to one-tenth of the reference frequency and meets the system stability guidelines, you do not need to consider the discrete-time domain. If you want to change it to about one-fifth or even closer, you must really consider the impact of sampling on dynamics, solve the sampling problem in the discrete-time domain environment, and ensure stability. Another consideration for PLL bandwidth is reference spurs. The narrow bandwidth will eliminate spurious signals through the filter to avoid such signals from entering the VCO. But the narrow bandwidth also requires the natural time domain response from these systems, which means that the response time is slower. Bandwidth has two considerations for noise. If you reduce the PLL bandwidth, you will get more VCO noise, but the reference and input noises will be more suppressed. If the bandwidth is enlarged, more VCO noises will be suppressed, but more reference and input noises will be generated. A larger bandwidth allows you to obtain a faster and better capture range without locking. Therefore, PLL bandwidth needs to be weighed according to the characteristics of different components.

3.2.2   Loop Optimization

Let us see how to optimize the loop to achieve the best phase-noise performance. The characteristics of the loop are the bandwidth ω_{-3dB} and damping factor ξ:

$$\omega_{-3dB} = \frac{I_{CP}RK_{VCO}}{2\pi N} \tag{3.1}$$

$$\zeta = \frac{R}{2}\sqrt{\frac{I_{CP}K_{VCO}C}{2\pi N}} \tag{3.2}$$

where I_{CP} denotes the charge-pump current; R, the loop-filter resistance; K_{VCO}, the VCO gain; and N, the frequency division.

We can select these parameters in different ways. For a given K_{VCO} and N, the charge-pump current and loop-filter components can be scaled with the same bandwidth and damping factor.

$$R \rightarrow \lambda R, I_{CP} \rightarrow \frac{I_{CP}}{\lambda}, C \rightarrow \frac{C}{\lambda} \tag{3.3}$$

Assuming another case, this is actually to emphasize the importance of having a low K_{VCO} in the transfer function where the noise amplification has been observed, but we can further quantify it here. So, suppose we have an oscillator gain K_{VCO}, its scale factor is β, the influence of K_{VCO}.

$$K_{VCO} \rightarrow \beta K_{VCO}, \quad I_{CP}R \rightarrow I_{CP}R/\beta \tag{3.4}$$

We can still keep the bandwidth and damping coefficient unchanged. Therefore, apart from a great compromise in terms of noise and system area, the behavior of the loop remains the same.

3.2.3   Trade-off among Phase Noise, Reference Spur, and Settling Time

Therefore, for a given PLL bandwidth, phase noise, reference spurs, and settling time can be defined. However, at the circuit level, we can still further improve

Fig. 3.8 Trade-offs

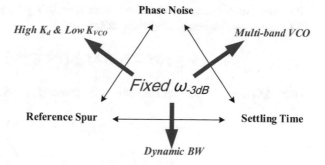

Fig. 3.9 Settling time

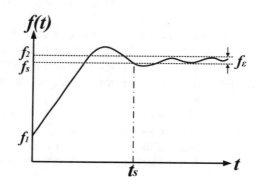

performance by breaking the trade-off between phase noise and reference spurs, as presented in Fig. 3.8. For example, even if we have the same PLL bandwidth, the PLL bandwidth can be obtained by multiplying the PD gain by the VCO gain, so we can assign a higher PD gain and a lower segment VCO gain so that we can have the same PLL bandwidth. The commonly employed method is the use of a multi-band VCO or DCO. While maintaining the same tuning range, the VCO gain will be significantly reduced, so the phase noise can be improved. In addition, a multi-band VCO can use a pre-tuned circuit (automatic frequency calibration), so the tracking time can be greatly shortened. Between the reference spur and the settling time, we can use the dynamic bandwidth, which means that we can increase the charge-pump gain during transients. Furthermore, when the PLL enters the lock period, we can set the charge-pump current to a normal value, so the reference spurious performance will not be affected.

3.2.4   Settling Time

Calculating the setup time of PLL is a complicated process, as presented in Fig. 3.9. In fact, we must rely on a MATLAB or PLL dedicated software. However, a simplified formula can be used to estimate the settling time. For second-order Type-2 PLL, transient frequency is approximated by

$$f(t) = f_1 + \Delta f_{\text{step}}\left(1 - e^{-\zeta\omega_n t}\right) \tag{3.5}$$

where $\Delta f_{\text{step}} = f_1 - f_2$, and the settling frequency f_s is within $f_\varepsilon = f_2 - f_s$.

$$f_s = f_1 + \Delta f_{\text{step}}\left(1 - e^{-\zeta\omega_n t_s}\right) \tag{3.6}$$

$$t_s = \frac{-1}{\zeta\omega_n} \ln\left(\frac{\zeta f_\varepsilon}{\Delta f_{\text{step}}}\right) \cong \frac{-2}{\omega_{-3\text{dB}}} \ln\left(\frac{\zeta f_\varepsilon}{\Delta f_{\text{step}}}\right) \tag{3.7}$$

The stabilization time is determined by the ratio of ζ, ω_n, the ratio of frequency step size Δf_{step}, and frequency error f_ε. Most of the current PLLs require automatic frequency calibration functions to adjust the VCO frequency in advance to overcome process and environmental changes while also significantly shortening the setup time.

Automatic frequency calibration is the foreground correction when the PLL is turned on. Since a multi-band VCO is used to reduce K_{VCO}, each sub-band needs to be separately calibrated during the calibration process so that the channel can be changed immediately during frequency hopping between sub-bands.

3.2.5   Reference Spurs

The reference spurs in the PLL are usually caused by the current loss in the charge-pump, the charge injection of the switch, and the charge sharing or leakage current in the LPF, as presented in Fig. 3.10. If the PLL has a static phase offset (generated internally by the phase detector or other mismatch), a DC offset voltage will be generated. However, as aforementioned, the PLL does not allow any frequency error; thus, the PLL must create another static phase error to compensate for the static phase error generated by the charge-pump. This is the reason why we see the voltage at the input of the VCO ripple. This operation is completed in each reference clock cycle, so the reference clock undergoes frequency modulation, and then the voltage ripple is determined according to the mismatch, which will determine the level of spurs. Frequency synthesizers used in some applications (such as wireless communications) have very strict requirements for spurs. By adding high-order poles, we can achieve additional spurious suppression.

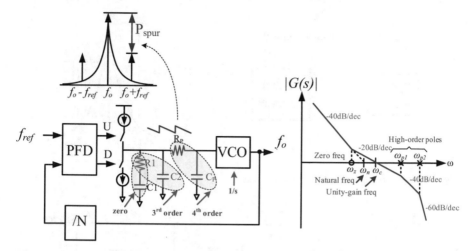

Fig. 3.10 Reference spur and spur suppression

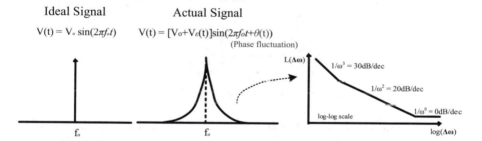

Fig. 3.11 Phase-noise curve

3.2.6 Phase Noise

By definition, phase noise is just a short-term phase fluctuation. The period of the cyclic signal cannot be kept constant for a long time, and this kind of frequency instability will cause phase noise. Ideally, a sinusoidal signal shows only one pulse in the frequency domain, but in reality, we always see phase fluctuations and some kind of phase noise near the carrier. Figure 3.11 presents the phase-noise curve. As can be seen from the figure, the phase-noise curve has three regions. The first is $1/\omega 0$, which is the thermal noise of the external clock added to the oscillator itself; it does not affect the oscillation time base. The second is $1/\omega 2$, which is the upconversion heat (AWGN) noise; it is caused by uncorrelated time fluctuations within the oscillation period, modeled as a random walk. The third one is $1/\omega 3$, which is the upconversion flicker (1/f) noise; it is caused by activation energy traps in thin MOS transistors.

Narrowband noise measurement

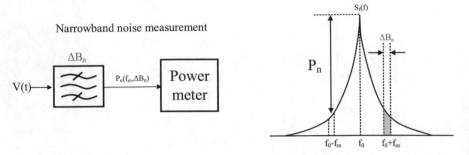

Fig. 3.12 Quantizing narrow-band phase noise

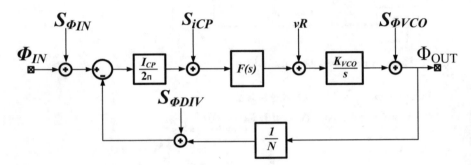

Fig. 3.13 Calculation of the noise transfer function

Phase noise is an important indicator of PLL. Because the phase-noise level of interest is much lower than the carrier power, we can apply narrowband FM theory to quantify the phase noise. First, noise is a random power, so the noise bandwidth must be defined. Once the noise bandwidth is defined, the phase noise can be quantified. The trick is to apply a very narrow noise bandwidth, as presented in Fig. 3.12. For each offset frequency, we define this small ΔBn. We want to measure the power in a narrow noise bandwidth. Second, the measured power is divided by the carrier power to obtain the normalized output. We generally use a noise bandwidth of 1 Hz, so the phase noise can be expressed in dBc, normalized to 1 Hz; so, we always use dBc/Hz to express the phase noise.

3.2.7 Phase Noise and Noise Transfer Function Calculation

We add all the phase-noise sources that need to be considered in the block diagram of the charge-pump PLL, as presented in Fig. 3.13, including the addition of input phase noise $S_{\Phi IN}$, charge-pump noise S_{iCP}, and loop-filter resistance noise S_{vR}, as well as the inherent phase noise $S_{\Phi VCO}$ of the VCO itself, and the noise of the divider $S_{\Phi DIV}$. Then, the overall phase noise of the output can be obtained.

The steps to calculate the total output phase noise are as follows. First, all noise sources ($S_i(f)$) must be identified, and then, the power spectral density of each noise source must be calculated or simulated. We must also evaluate the transfer function from each of the noise source to the output, ($H_i(f)$). Once we know it, all we have to do is take each of these power spectral densities and multiply it by the power gain from that point to the output, i.e., $|H_i(f)|^2 S_i(f)$. Then, add all of them together, i.e., $\sum |H_i(f)|^2 \cdot S_i(f)$.

If we only need to see the contribution of a single specific thermal noise, such as the contribution of the VCO or the contribution of the loop filter's thermal noise, we can pass the noise transfer function (NTF) from the VCO or loop filter to the output and then obtain the result, i.e., $|NTF(s)|^2$. Subsequently, multiply that by the power spectral density of the noise (S_{noise}); finally, add all these terms. Since the noise of the frequency divider is very small, it is usually ignored.

The power spectral density of each noise source is as follows:

$$S_{\Phi_{OUT}}^{\Phi_{IN}} = S_{\Phi_{IN}} \cdot |NTF_{IN}(s)|^2 \tag{3.8}$$

$$S_{\Phi_{OUT}}^{i_{CP}} = S_{i_{CP}} \cdot |NTF_{CP}(s)|^2 \tag{3.9}$$

$$S_{\Phi_{OUT}}^{V_R} = S_{V_R} \cdot |NTF_R(s)|^2 \tag{3.10}$$

$$S_{\Phi_{OUT}}^{\Phi_{VCO}} = S_{\Phi_{VCO}} \cdot |NTF_{VCO}(s)|^2 \tag{3.11}$$

Noise Transfer Functions [2]

$$NTF_{IN}(s) = \frac{\Phi_{OUT}(s)}{\Phi_{IN}(s)} = N\frac{2\zeta\omega_n s + \omega_n^2}{s^2 + 2\zeta\omega_n s + \omega_n^2} \tag{3.12}$$

$$NTF_{CP}(s) = \frac{\Phi_{OUT}(s)}{IN(s)} = \frac{(K_{VCO}/C)(1 + sRC)}{s^2 + 2\zeta\omega_n s + \omega_n^2} \tag{3.13}$$

$$NTF_R(s) = \frac{\Phi_{OUT}(s)}{V_n(s)} = \frac{sK_{VCO}}{s^2 + 2\zeta\omega_n s + \omega_n^2} \tag{3.14}$$

$$NTF_{VCO}(s) = \frac{\Phi_{OUT}(s)}{\Phi_{VCO}(s)} = \frac{s^2}{s^2 + 2\zeta\omega_n s + \omega_n^2} \tag{3.15}$$

It can also be observed that each transfer function from input to output has different characteristics. The output charge-pump current noise is a low-pass transfer function, the resistance noise from the loop filter to the output is a band-pass transfer function, and the output VCO noise is a high-pass transfer function. It is worth noting the charge-pump noise and loop-filter noise, both of which include K_{VCO} in the numerator of the transfer function. This means that these noise sources will be amplified by the oscillator gain, and their influence should be considered when designing the K_{VCO}.

3.3   Building Blocks of Integer-N Frequency Synthesizer

There are five building blocks in an integer-N frequency synthesizer, and each building block has many different aspects of circuit design. Therefore, some important concepts and nonideal aspects of each building block will be briefly introduced as follows.

3.3.1   Phase Detector

The phase detector can detect the phase difference between the input and the feedback input to create a pulse output waveform, where the pulse width corresponds to the actual phase difference signal. Figure 3.14 presents a simple phase detector, such as XOR, or multiplier phase detector (for high frequency only).

This operation can also be done very easily using standard digital gates. The basic idea is that when the edges are basically separated from each other, the corresponding pulse width also changes. It also performs frequency acquisition to a certain extent.

If the pulse signal output by the phase detector is averaged, the averaged signal corresponds to a very smooth linear characteristic curve with phase error. The

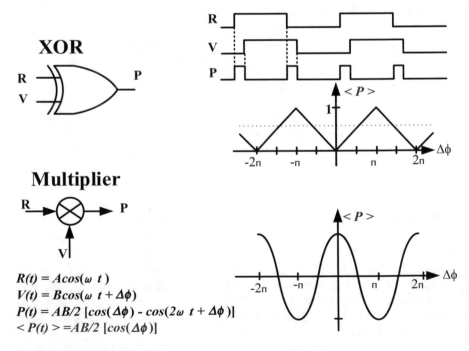

$$R(t) = A cos(\omega\ t\ )$$
$$V(t) = B cos(\omega\ t + \Delta\phi)$$
$$P(t) = AB/2\ [cos(\Delta\phi) - cos(2\omega\ t + \Delta\phi\ )]$$
$$< P(t) > = AB/2\ [cos(\Delta\phi\ )]$$

Fig. 3.14 Simple phase detector

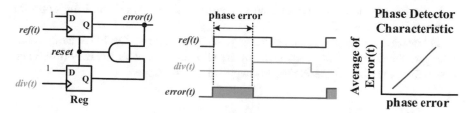

Fig. 3.15 Phase detector implemented by simple gates and its characteristics

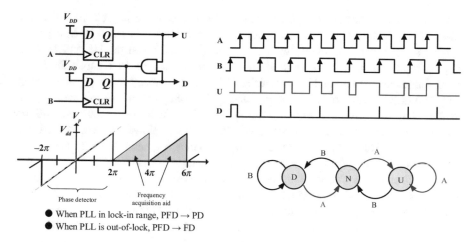

● When PLL in lock-in range, PFD → PD
● When PLL is out-of-lock, PFD → FD

Fig. 3.16 PFD function and state diagram

concept of averaging requires a very good analog filter to complete this operation, as presented in Fig. 3.15.

Figure 3.16 is the block diagram of the PFD. This is a very simple block diagram and is just a few flip-flops with asynchronous reset function. It functions as follows: if the reference signal is ahead of the VCO signal, the U (UP) signal will appear first. Similarly, if the feedback signal is input first, the D (DOWN) signal will appear first, indicating that the speed of the oscillator must be reduced.

The transfer function of the phase detector is presented here. For a positive phase difference, a positive output will be obtained, and for a negative phase difference, a negative output will be obtained. It can be seen that the linear range is from -2π to $+2\pi$, so the gain of the phase detector is actually $V_{DD}/2\pi$.

When the PLL enters the lock range, the PFD is just a traditional linear phase detector. Its difference from other phase detectors is only the linear range and the gain of the PD, but when the PLL exceeds the lock range, it means that some cycle slips can be observed. Thus, some helpful information for frequency acquisition is provided here; refer to the time domain example. The 2π radians only work in the bipolar direction (positive and negative), but when the phase error exceeds the 2π radians, as we have memory (state machine), the memory retains the original unipolar information. Positive charge is continuously provided until the opposite message is encountered.

For charge-pumps, PFD is not necessary. We can still use XOR PD or a multiplier to trigger the charge-pump. Because they are all phase detectors, the only difference is that the PD gain and the PD linear range are not the same.

3.3.1.1   Nonideal PFD: Dead Zone

As aforementioned, the PFD generates a pulse whose width is proportional to the phase error. When the zero phase error or the phase error is small or in a phase-locked state, a very narrow pulse will still be obtained when the UP and DOWN signals are aligned. If the pulse is too narrow and the output drive capability is not sufficient to drive the metal wiring of the MOS switch capacitor and gate capacitor, the PFD output may not reach the power-supply voltage or rail-to-rail swing, so it will not be transferred to the charge-pump. The PD gain is greatly reduced, and the so-called dead zone appears, as presented in Fig. 3.17. However, even if we get a strong output from a PFD (an inverter that provides a greater drive capacity), the charge-pump itself will take some time to generate the same number of charge-pumps. The simple solution is to add a delay circuit to the reset path to solve all problems. This is because when the phase error is zero, the common delay line provides the same pulse width as it provides sufficient TD time to fully turn on the charge-pump. Moreover, the PLL ultimately only considers the difference between the rising current and the falling current. Therefore, even if the minimum pulse width is used, as long as these pulse widths are the same, there will be no net change in the charge of the loop filter.

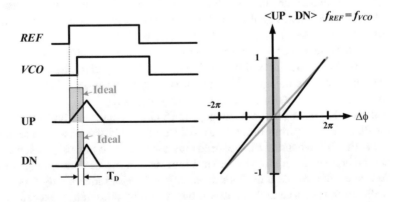

Fig. 3.17 Nonideal PFD: Dead Zone

3.3.2   Charge-Pump

We need to convert the output of the PFD into an analog voltage. It can be done in voltage mode or current mode. If it runs in the current mode, it will have certain advantages. For Type-2 behavior, we need an integrator, as presented in Fig. 3.18. We can also realize this integrator function through charge-pumps and passive components. By adding an extra resistor, an extra pull-up zero is actually added, which can increase the stability of the PLL.

To illustrate the function of the charge-pump, let us start with this familiar active lag-lead filter. When we have an UP signal, the upper switch is open, and a positive voltage is connected to the node where the resistor is connected to the virtual ground. For the DN signal, we will get a voltage of the opposite polarity. Now, if we divide the voltage by the resistance and replace the current value with one current source, then we can get the same function by using two current sources. Therefore, the passive loop filter can now perform the same function as the active loop filter.

Now, we are going to analyze the transfer function of the charge-pump, assuming that there is no resistance. In Fig. 3.19, we also show that the reference signal leads the VCO feedback signal with an error of Φ_e, so the net UP pulse is wider than the DN pulse Φ_e, which will cause a certain charge to dump on the loop-filter capacitor. The magnitude of the charge is consistent with the phase difference. Therefore, the average charge dump per cycle is actually $I_{cp}/2\pi \times$ phase error.

The average charge-dump per cycle is as follows:

$$\overline{Q_{ctrl}} = \frac{C_1 \Delta V_{ctrl}}{T} = \frac{I_{CP}}{2\pi} \cdot \Phi_e \qquad (3.16)$$

Fig. 3.18 Charge pump and loop filter

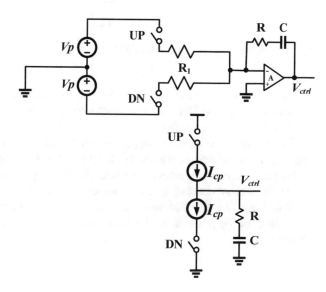

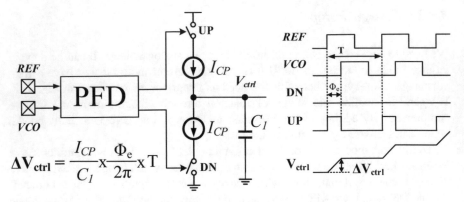

$$\Delta V_{ctrl} = \frac{I_{CP}}{C_1} \times \frac{\Phi_e}{2\pi} \times T$$

Fig. 3.19 Working principle of the charge-pump

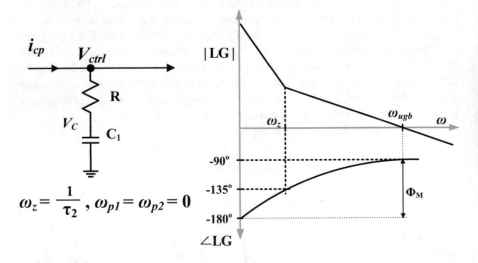

$$\omega_z = \frac{1}{\tau_2} \ , \ \omega_{p1} = \omega_{p2} = 0$$

Fig. 3.20 Stability of the charge-pump PLL

The combination of PFD and charge-pump is modeled by $I_{cp}/2\pi$ [ampere/radian]. We can represent the PFD and the charge-pump together as a block with a transfer function $I_{cp}/2\pi$ in ampere/radian.

For stability, we need to connect a resistor in series with C1, as presented in Fig. 3.20. The reason is that if the loop transfer function is used, the loop gain without resistance will decrease at a rate of −40 dB per decade, and the phase lag is −180°. So, before the loop gain reaches unity gain, zero needs to be added to reduce the lag. Then, the resistor provides the zero needed to make the phase rise, and depending on the position of the zero point, a phase margin of almost 90° can be obtained.

Phase margin:

$$\Phi_M = \tan^{-1}\left(\frac{\omega_{ugb}}{\omega_z}\right) \tag{3.17}$$

Loop Gain:

$$LG(s) = K_{pd} \cdot \frac{1 + s\tau_2}{s\tau_1} \cdot \frac{K_{VCO}}{s} \tag{3.18}$$

In Fig. 3.21, the transient waveform of the charge-pump charging the capacitor can be observed after adding a resistor. The voltage we see on the capacitor will accumulate all past error phase information. This is the integrated path that provides frequency tracking information. In addition, the voltage across the resistor provides a proportional gain path. Therefore, the VCO input contains two voltages: one is a small signal, which shows our instantaneous phase tracking, and the other is a large signal, which contains frequency acquisition information. When we use a charge-pump current of 1 mA and a resistance of 10 k, the instantaneous voltage across the resistance can reach 10 V, which is higher than the power-supply voltage. As a result, as long as we have this kind of jump, it will saturate the output stage of the charge-pump and the input stage of the VCO. Therefore, to protect the charge-pump and VCO, parallel capacitors must also be placed.

The addition of parallel capacitors in Fig. 3.22 has another effect. Even if the loop is locked, there should be periodic rising and falling pulses. In theory, these pulses should cancel each other out, but due to some undesirable characteristics, the loop-filter voltage will periodically experience a small offset and cause interference. Any periodic interference on the node will produce tones, which are called reference spurs. An easy way to eliminate them is to simply add a small capacitor C2 in the loop filter. Although this will slightly reduce the phase margin, it will improve the ripple.

Fig. 3.21 Interpretation of the CP transient waveform

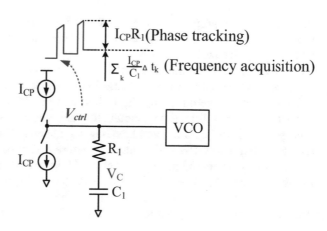

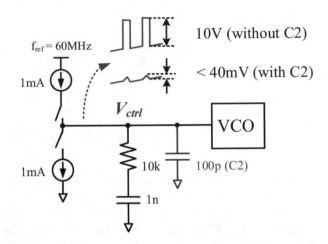

Fig. 3.22 Charge-pump ripple bypass

3.3.2.1 Nonideal Charge-Pump

The performance of the charge-pump will be affected by the nanotechnology operating under lower voltage. Because the low supply voltage limits the output swing of the charge-pump, the oscillator is forced to obtain higher gains. In addition, the low supply voltage reduces $V_{GS} - V_T$, which can easily lead to poor current mirror matching and causes mismatch of the upper and lower currents of the charge-pump. Since the output conductance of short-channel devices is not that large, when we deviate from the ideal center point of the loop-filter voltage, it will also cause current mismatch. Subthreshold leakage and temperature changes will aggravate the current mismatch. Gate leakage can also exacerbate current mismatch. Combining the above various current mismatch factors will lead to increased spurs. The following explains some of the nonideal conditions in the charge-pump:

3.3.2.2 Nonideal Charge-Pump: Current Mismatch

In a nonideal charge-pump, if there is an up/down current mismatch, a phase shift will occur in the digital clock generation, or a reference spur will be generated in the frequency synthesizer.

Theoretically, when the phase of the PLL is locked, the up/down signals are balanced, and we cannot see any change in the output of the charge-pump. Since the rising current is determined by the PMOS and the falling current by the NMOS current source, it is difficult to maintain the exact same bias in different controllable voltage ranges. In Fig. 3.23, we show how the mismatch between the rising current and the falling current actually causes the reference spurs, assuming that the rising current is lower than the falling current by ΔI. Moreover, when the loop is locked, the rising pulse (T_{ON}) will be smaller than the falling pulse (T_{rst}) to balance the charge. In addition, the PLL does not allow any DC offset voltage to produce any frequency error, and at each reference clock cycle, the accumulated error is corrected

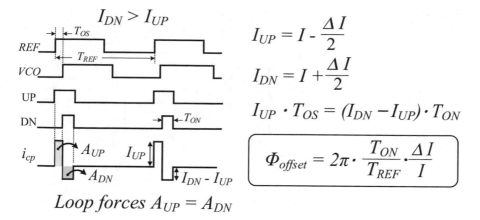

$$I_{UP} = I - \frac{\Delta I}{2}$$

$$I_{DN} = I + \frac{\Delta I}{2}$$

$$I_{UP} \cdot T_{OS} = (I_{DN} - I_{UP}) \cdot T_{ON}$$

$$\Phi_{offset} = 2\pi \cdot \frac{T_{ON}}{T_{REF}} \cdot \frac{\Delta I}{I}$$

Fig. 3.23 Charge-pump current mismatch

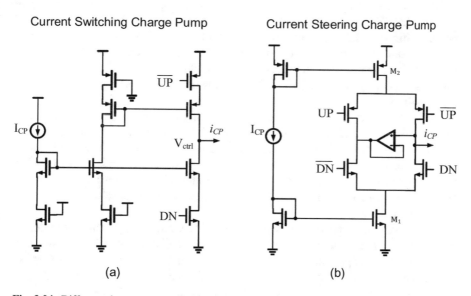

Fig. 3.24 Different charge-pump switching schemes

together. This is a result of static phase error caused by current mismatch. The loop will stabilize at an offset voltage proportional to the current mismatch. Therefore, the offset will cause periodic disturbances in the loop-filter voltage and then spurs.

Even for the same charge-pump mismatch, if the on-time of the PFD is reduced, the spurious level can be reduced. However, the higher the frequency division ratio, the more the reference spurs will increase. Thus, when designing a PLL with a high-frequency division ratio, the mismatch requirements of the charge-pump will become larger and more difficult to meet.

Fig. 3.25 Reduced CLM and mismatch

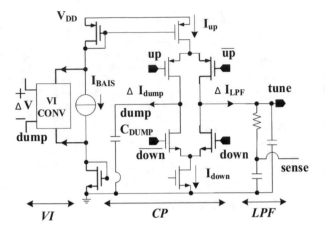

The current-switching charge-pump presented in Fig. 3.24(a) was popular some time ago [3]. The charge-pump uses a source-side switch, so when the falling signal becomes high, the falling current will be drawn from the filter through the switching transistor and the current mirror transistor. Similarly, when the upstream signal becomes low, the current that needs to be supplemented will flow into the loop filter, so we can convert the UP and DN signals into electric charge and then dump them to the filter.

Due to the need to use smaller devices, resulting in lower output impedance and higher leakage current, or due to channel length modulation, the falling current of NMOS will no longer completely match the rising current from the PMOS, which will lead to poor matching. The mismatch between the current mirrors will lead to the mismatch between the charge pump rising and falling current. This can cause static phase errors, which in turn can lead to reference spurs. Therefore, this is not an attractive candidate.

Figure 3.24(b) demonstrates that the current steering charge-pump is a current control rather than a current switch [4]. The two transistors, namely, M1 and M2, which indicate the upper and lower tail current sources, are not switched, so they can be lengthened to make their matching and output conductance better. The switching device can be made smaller, so that it can be switched faster. Moreover, the actual output node of the charge-pump can connect the output copy to the virtual node through a unity gain buffer, so the second-order effect of charge sharing can be reduced. This is a very good architecture that is still in use today.

Figure 3.25 presents an example of a charge-pump that solves not only the problem of mismatch but also the problem of channel length modulation (CLM) [5]. The right illustration in Fig. 3.25 is a traditional charge-pump, and that on the left is the VI converter that can correct the mismatch. When locked, the PLL will converge to a state where the average current flowing into the loop filter is zero. In the case of charge-pump rising/falling current mismatch, the PLL shifts the VCO phase relative to the reference phase to maintain charge balance. Although the tuning voltage may be locked, C_{DUMP} will continue to increase or decrease based on the

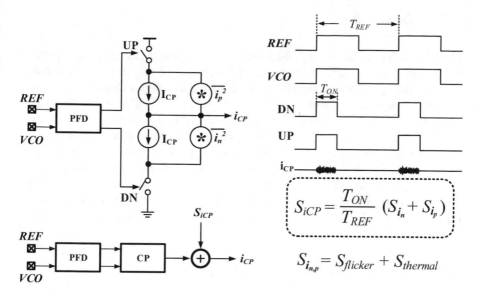

Fig. 3.26 Charge-pump noise

signs of mismatch. The VI converter compares the voltage between the C_{DUMP} loop filter and the loop filter and adjusts the rising and falling currents until the two voltages are equal and stable. This means that the rising and falling currents of the charge-pump match in a steady state.

The power spectral density analysis of the charge-pump noise is explained as presented in Fig. 3.26. When the PLL reference phase and VCO feedback phase will be locked to the near-zero phase error, the rising and falling signals will be turned on for a specific time. When these two currents are turned on, the current will pass through the transistor, but the generated noise will enter the loop filter. During the conduction period of the two current sources, the noise can be calculated as the sum of the power spectral densities of the two noises, $S_{in} + S_{ip}$, including the flicker and thermal noises. Because the passing current is usually fixed, to reduce the noise of the current source, what we can do when designing the circuit is to make these g_m/I_D as small as possible. This means using a larger $V_{GD} - V_T$, at the cost of leaving more head-room on the transistor.

Since this noise is cyclostationary noise, these expressions in Fig. 3.26 are only valid when the current source is turned on. The dynamic characteristic of PLL is much slower than the reference frequency. Therefore, we can actually eliminate the noise by averaging it and then writing its power spectral density T_{ON} multiplied by T_{REF} multiplied by the noise ratio when both are turned on, as presented in Fig. 3.26. Therefore, minimizing the on-time also helps reduce the noise impact of the charge-pump.

3.3.3 Loop Filter

The loop filter converts the digital output of the discriminator into an analog voltage. As presented in Fig. 3.27, it has a pole at the origin and zero to maintain stability. High-frequency poles are usually introduced to reduce spurious tones.

The loop filter can be implemented with an active filter, but the problem with the active filter is that the operational amplifier generates more noise and consumes more power. The passive loop filter generates less noise and better power-supply rejection performance. Therefore, passive filters are indeed the first choice in many designs.

Generally, Fig. 3.27 presents an active filter configuration. Also called lag-lead filter, the transfer function is $(1 + s\tau2)/s\tau1$. Therefore, it can be written as an integer term and a proportional term. In terms of signal flow, there can be two paths, namely, the integration path and the proportional path, and finally the two paths are added.

3.3.3.1 Problems with Analog Loop Filters

Because it is a passive loop filter, the only components are resistors and capacitors. For resistors, we can maintain the original bandwidth and damping by adjusting the ratio of the resistor value to the charge-pump current. For capacitors (usually damping capacitors), the value is between 10 and 100 pF. As the purpose of the capacitor is to provide sufficient damping or phase margin, it does not have to be linear. Moreover, because the capacitance is proportional to the area of the device and MOS capacitors provide the highest capacitance density, we usually use MOS to build these large capacitors. Thus, in the past, MOS capacitors have been used up to the 130-nm node. But in the technology below the 90-nm node, gate leakage is becoming a problem, because it will also be proportional to the area of the device. As presented in Fig. 3.28, under the influence of the gate leakage current, the DC voltage level will change, so the PLL must provide a phase offset to eliminate

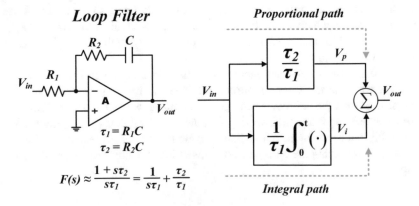

Fig. 3.27 Loop filter

Fig. 3.28 MOS capacitor
leakage in the loop filter

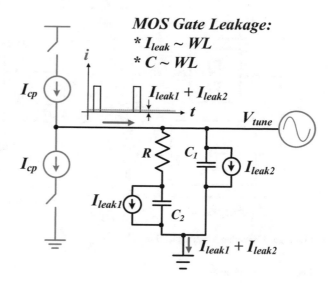

it. As aforementioned, for each reference clock cycle, the net balance of charge should be equal. Similar to the charge-pump current mismatch, it increases the amount of jitter and increases the spurious level.

For capacitance leakage compensation, [6] proposed a compensation scheme. The idea is to first match the leakage current with another similar device. The circuit implementation is quite complicated, and an additional noise source is introduced simultaneously. The impact of noise must be evaluated very carefully. Other different methods can also be evaluated to avoid the abovementioned problem. If all other methods fail, a thick oxide device or a higher power-supply voltage can be used. Because thick oxide devices are used in the I/O unit, they will not produce any leakage. But the problem is that they occupy a large area.

Suppose we have a 90-nm design, and we want to port it to 45 nm. If the newly designed capacitor is larger than the old one, we can consider a capacitor multiplication scheme.

Capacitance multiplication schemes have existed for many years. The working principle of all these schemes is summarized below. There is a transfer function $I * R$ on the proportional path and a transfer function I/C on the integral path. Now, to reduce C, all we have to do is reduce I so that the I/C ratio can remain the same. However, if we do this, to keep $I * R$ constant, we must first increase the resistance and then reduce the overall phase noise. Another method is to separate the proportional path and the integral path. Referring to Fig. 3.29 [7], we have a current I flowing through the resistor R and then a current αI, where $\alpha < 1$. This flows through the capacitor. Then, the two voltages generated here must be added; usually, an operational amplifier needs to be added. However, the operational amplifier will add noise to the entire system, introduce more frequency poles, affect the frequency response of the PLL, and consume more power.

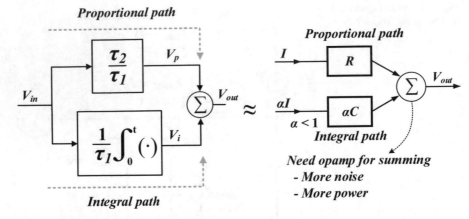

Fig. 3.29 Capacitance multiplication [7]

Fig. 3.30 Capacitance
multiplication [8]

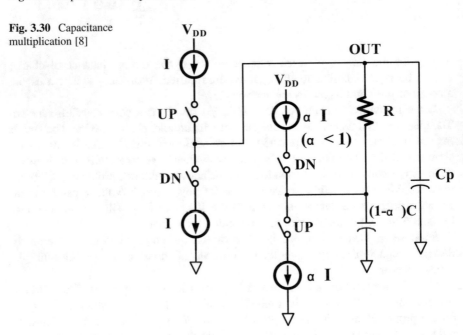

Another example is presented in Fig. 3.30 [8]. We still continue to use passive loop filters, as presented in the figure below. Furthermore, we do not use op amps, but instead two charge-pumps. We have a main charge-pump with a current of I and then an auxiliary charge-pump with a current of $\alpha * I$. Now, α is less than 1. Please pay attention to the switching phase here. When the rising signal is turned on and the main charge-pump pushes the current into the resistor, the auxiliary charge-pump draws current from the node between the resistor and the capacitor. Therefore, the result is that only $1-\alpha$ times the current will flow to the capacitor. Therefore, we can set the damping capacitor from C to $(1-\alpha) * C$. The equivalent capacitor value is amplified as $C/(1-\alpha)$.

3.3.4   Ring and LC Oscillators

For a single-chip ring oscillator, we can choose the ring oscillator or the on-chip LC oscillator.

3.3.4.1   LC Voltage-Controlled Oscillator

The advantage of the LC oscillator is that it has a good phase noise and high oscillation frequency and is not affected by process and temperature changes. The disadvantage is that the tuning range is limited, and the occupied area is large. The inductance value is actually determined by the inductance area required to build it. In addition, it only provides a limited number of clock phases. Therefore, the benefit of technology scaling to the LC oscillator is not obvious, as the oscillation frequency is determined by the inductance and capacitance on the chip; thus, the smaller the area of the inductance and capacitance, the higher the frequency that can be obtained. Therefore, a possible method is to use a smaller LC oscillator area, work at a higher frequency, and then use a frequency divider to obtain the same frequency.

Figure 3.31 presents the inductor implemented on the chip. If we want to implement a resonator, we need to match the capacitors and connect them in parallel to the differential inductor, the center-tapped inductor, and the two tank capacitors. The electrical characteristics are what we care about and will be discussed. For ease of description, it is assumed that the parasitic effects of inductance and capacitance are sufficiently small.

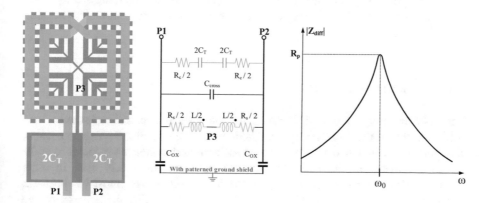

Fig. 3.31 On-chip LC resonator

Assuming $C_T \gg C_{OX}$ and $C_T \gg C_{cross}$, then we get an LC resonator, and the resonance frequency and inherent impedance are as follows:

$$\omega_0 \approx \frac{1}{\sqrt{L \cdot C_T}} \tag{3.19}$$

$$Z_0 = \sqrt{\frac{L}{C_T}} \tag{3.20}$$

Because the topology in Fig. 3.31 is parallel resonance, the impedance will reach a peak near the resonance frequency. At resonance, we can obtain the characteristic impedance, which is equivalent to the resistance impedance of the tank.

$$R_P = Q_T \cdot Z_0 \tag{3.21}$$

The quality of the oscillator is determined by the quality factor, which represents the stability of the oscillator. The quality factor of a resonator is a combination of the quality factors of a capacitor and an inductor.

The inductance, capacitance, and total Q values are as follows:

$$Q_L = \frac{\omega L}{R_S} \tag{3.22}$$

$$Q_C = \frac{1}{\omega C_T \cdot R_C} \tag{3.23}$$

$$Q_T = Q_L \parallel Q_C \tag{3.24}$$

In fact, if we look at this formula, the smallest quality factor will dominate the quality factor of the entire tank. The quality factor of an inductor tends to increase with the increase in frequency. The quality factor of a capacitor will decrease with frequency. Therefore, in low-frequency applications, the quality factor will be limited by the quality of inductor. For high-frequency applications, the quality factor of the capacitor may be an issue, or when we have a widely adjustable oscillator, both are issues.

3.3.4.2 Varactors in CMOS

The resonant frequency is determined by the capacitance and inductance, and adjusting any of them can change the frequency of the oscillator. This is because changing the capacitance of the varactor is easier than changing the inductance of the on-chip inductor. In the past, PN varactor diodes were widely used in older manufacturing processes. However, in the MOS technology, most of the time, people tend to use MOS-based varactor. As presented in Fig. 3.32, we have two options, using reverse varactor diodes or using accumulative varactor.

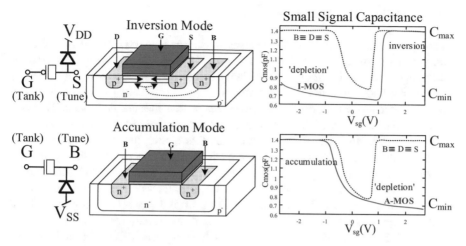

Fig. 3.32 Varactors in CMOS [9]

The inversion-mode varactor is nothing but a PMOS transistor. In this transistor, we connect a source and a drain together, and we get a two-terminal device between gate and source. If we make the gate voltage much smaller than the source voltage, the PMOS transistor will form an inversion layer, holes out of the source and drain will go under the gate plate, and we will get a high inversion-mode capacitance. If we now make the gate voltage larger than the source voltage, the structure would like to go into accumulation-mode. However, this structure cannot do that because it does not have the source of electrons. So, it stays in depletion and we get a small capacitance under this condition. Therefore, we can see that by switching the voltage across the varactor diode, the capacitor can be switched between a lower value and a higher value, and a good ratio of C_{MIN} to C_{MAX} can be obtained.

The accumulation-mode varactor is also a PMOS transistor, but has a different structure. The original p + source just needs to be replaced and drained with n + implantation. When the source voltage is greater than the gate voltage, there is no available inversion charging source, and the structure is actually in a "depleted" state. However, if we bias the gate at high voltage and the source at low voltage, the electrons from this n + source will enter the area under the gate and form an accumulation layer, and we get high capacitance again. Therefore, we can switch between the C_{MIN} and C_{MAX} values again.

In the actual usage, when we use these varactors, we find that parasitic effects are very important, and there are many parasitic effects on the source. Normally, we connect the gate side to the tank and use the source side to connect the tune nodes, as this will reduce the load on the tank.

3.3.4.3   Adjustment Range

In many wireless applications, if the carrier frequency is in the range of 1–10 GHz, a tuning range or frequency variation of 20–200 MHz is needed. When converted to percentage, the tuning range of the relative range $\Delta f/f_0$ is approximately $\pm 1\%$ to $\pm 2.5\%$. We know that we are facing process variation. Usually, process changes do not have much effect on the inductance, because the inductance is actually determined by the geometry of the structure, and the geometry of the structure is usually well controlled. However, the capacitor is another matter, because in the capacitor, we can easily see the process variation of $\Delta C/C \cong \pm 20\%$. Because of the square root correlation between capacitance and frequency, we actually found that due to process changes, a frequency change of $\pm 10\%$, which is $\Delta f/f_0 \cong \pm 10\%$, can be expected. In fact, the tunability of the VCO we need depends, to a large extent, on the overcoming of PVT changes rather than application requirements.

We see that as the V_{DD} drops in the nano process, the VCO gain tends to rise, which will cause larger reference spurs and also cause the VCO to generate more noise. The common method is to use a discrete range with digital adjustment mechanism. As presented in Fig. 3.33, we can switch the adjustment elements (inductor or capacitor) to switch between these characteristics, so that the overlapping tuning characteristics cover the continuous tuning range.

Figure 3.34 presents the three types of oscillator bias topology: (1) Tail bias topology. Basically, the high common-mode level of the signal will be close to the supply voltage. The swing of this oscillator will exceed the V_{DD}. In this way, the V_{GD} voltage will reach twice the power supply. This can cause a considerable amount of stress on the gate and drain of the device, which may generate reliability problems. Likewise, the high common-mode level of the signal will also reduce the flexibility of obtaining a good tuning range in the varactor. (2) Top bias topology. In this case, we can keep the common-mode level of the signal at about half of the power supply and ensure that all signal swings are within the power rail. Of course, the signal swing will be smaller now, although the noise is worse than (1), but in many cases, we have better reliability. In addition, since the half-supply common mode is usually more compatible with varactor biasing, the adjustability will be better. (3) Complementary topology. Under the same bias, i.e., (1), since the inverter pair is composed

Fig. 3.33 Overlap tuning characteristics with wide adjustment range [10]

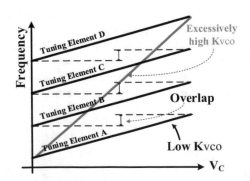

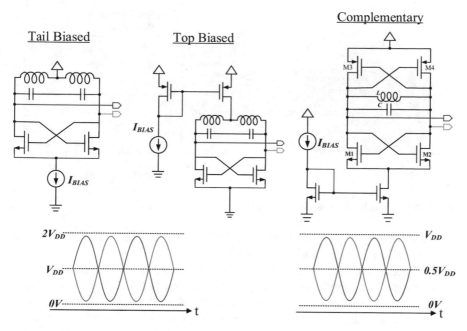

Fig. 3.34 Tail bias, top bias, or complementary VCO

of the PMOS and NMOS transistors, they tend to provide greater swing and slew rate.

3.3.4.4 Stacked MOS Capacitors and Inductors

Because the area occupied by the inductor is too large, how to make full use of the space under the inductor is a very important question. A capacitor needs to be created in the area below the inductor in Fig. 3.35 and used for the loop filter. This capacitor is a MOS capacitor [11, 12]. The layout needs to be created carefully so that there is no eddy current path under the inductor. The paper proves that the quality factor is not degraded due to structural changes.

3.3.4.5 Ring Voltage-Controlled Oscillator

The advantage of the ring oscillator is that it is very compact and has a wide tuning range. It can have multiple clock phases simultaneously, which is very useful for certain communication protocols or applications. The area of the oscillator is directly scaled by technology, and the oscillation frequency is not as high as the LC oscillator. However, the increase in the scaling ratio helps in the generation of a

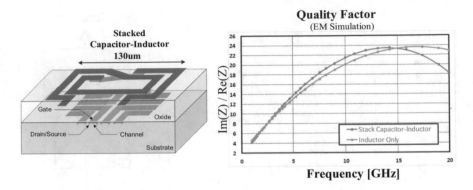

Fig. 3.35 Stacked MOS capacitor and inductor

higher oscillation frequency. The phase noise is higher than the LC oscillator, and it is very sensitive to process and voltage changes.

The ring oscillator is an oscillating circuit composed of a series of delay units. Assuming that the signal introduces a certain phase shift after each delay if the total phase shift around the loop is an integer multiple of 360°, the loop will oscillate. If we have single-ended inverted delays, then we need an odd number of delays, with a minimum of 3. For differential delay, we can have even or odd delay stages. The oscillation period does not exceed twice the number of delay elements multiplied by the delay τ of each stage, and the phase interval is 180°/N. For some communication applications, we need quadrature clock phases, so an even number of delay elements, such as 2 or 4, are required.

3.3.4.6   Requirements and Technology Impact of Ring Oscillators

As aforementioned, the ring oscillator must have a wide adjustment range and linear tuning characteristics within a wide operating frequency range. It means that the loop bandwidth and phase margin need to be effective within the loop dynamics of the wide operating frequency of the PLL. Moreover, special attention should be paid to the power-supply rejection ratio. Otherwise, phase noise or PLL jitter will be degraded. The design of the ring oscillator is influenced by the low operating voltage of nanotechnology in the following aspects, and the gain of the oscillator must be increased to cover a certain tuning range. Like LC VCO, on the SoC, because the noise (d_i/d_t) of the power supply and substrate is sometimes very high, if such a noise is coupled to the VCO input in some way, it will be amplified to the output VCO, and then the entire phase noise is degraded. Similarly, if the power supply is handled carelessly, it may reduce the phase noise of the entire oscillator.

3.3.4.7   Ring Oscillator Topology

Figure 3.36 presents some oscillator topologies. An oscillator can be divided into small and large signals as follows.

The small-signal oscillator can be a simple common source, as presented in Fig. 3.36(a). It is a series of common-source amplifiers. Its swing is very small. In the large-signal oscillator with a series of inverters presented in Fig. 3.36(b), the output swings from rail to rail, so the swing is high. However, the power-supply rejection ratio is very poor, and a voltage regulator must thus be used to power it. Furthermore, its phase noise is very good, and the power consumption is very low. An oscillator can also be categorized into single-ended (Fig. 3.36(c)) and differential (Fig. 3.36(d)). If we want to obtain an even number of phases and an even number of delay stages in the oscillator, just invert the signal in the feedback.

Figure 3.36(e) presents a differential small-signal delay unit, which was very popular in the past. It is a common-source stage of triode devices, with a good PSRR (linear load) rate, and its output signal swing is small, resulting in poor phase noise. The tail current itself is always turned on, requiring more power and always producing noise.

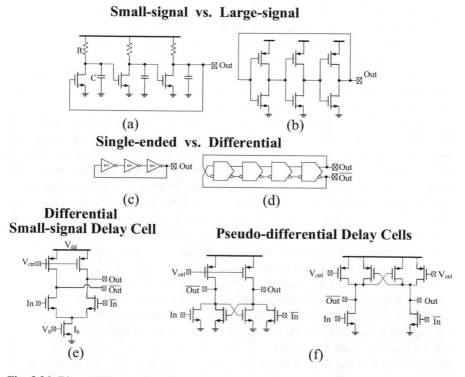

Fig. 3.36 Ring oscillator topologies: (**a**) Small-signal, (**b**) Large-signal, (**c**) Single-ended, (**d**) Differential, (**e**) Differential small-signal delay cell, and (**f**) Pseudo-differential delay cells

In addition to using inverters, the delay cell can also use pseudo-differential delay cells, as presented in Fig. 3.36(f). It has two possible implementation topologies, namely, NMOS cross-coupling and PMOS cross-coupling. The adjustment range of these delay units is very wide, but the rise and fall times are asymmetrical. The duty cycle distortion caused by this asymmetrical output waveform leads to 1/f noise upconversion, which will degrade the phase noise of the oscillator, and the power-supply noise suppression ability is poor, and it is as sensitive to V_{DD} and V_{ctrl}, which can be overcome by using a voltage regulator.

3.3.4.8   VCO Design Considerations

VCO design considerations are summarized as follows: (1) Quality factor (inductance or capacitance): it affects phase noise and swing. (2) The maximum output swing of the VCO: the output swing can reach the ohmic region of some devices before saturation. A higher output swing means that for a given noise, we will get a greater SNR, so the phase noise will be better. (3) Amplitude control: phase noise will vary with frequency changes or changes between components. Therefore, there must be a mechanism to control the output swing so as not to enter the bad operation region, which causes a low swing. (4) Within the tuning range, a low and constant K_{VCO} must be maintained. This is because the spurious power is proportional to K_{VCO}. If the K_{VCO} is changed, the loop equation and stability of the PLL may also be changed. (5) Since the VCO is very sensitive to power-supply noise, we hope to keep the power-supply noise at a very low level. It is important to have a good voltage regulator to isolate the noise generated by the VCO from the power supply. (6) Finally, as we have seen before, asymmetrical waveforms can cause 1/f noise, which can be upconverted or downconverted.

The characteristics of the power-supply voltage will affect the performance of the VCO. Not only the oscillator can be tuned by the tuning node, but most oscillators can also be tuned by the power-supply voltage. In the example in Fig. 3.37, the power-supply voltage sets the common mode of the signal. Although we keep the tuning voltage constant, changing the power-supply voltage will also change the voltage across the varactor, which results in a change in the oscillator frequency. This means that if we interfere with the power supply, phase noise will be generated, or a frequency modulation effect will be induced, which will become phase noise. Therefore, in many oscillators, especially those that must be placed on the SoC, it is important to use a good PSRR to ensure that they can exhibit excellent performance.

If we put everything on the chip, as indicated by the left illustration in Fig. 3.38, there is a voltage regulator to power the VCO and charge-pump. Even if we are very careful in the layout, there will be a spatial distance between the charge-pump, power supply, and VCO as well as an impedance. This means that when the charge-pump starts to pull the pulse current, we will obtain the power-supply ripple of the VCO, which will possibly generate reference spurs. As can be seen from the right figure, if we use another mechanism of off-chip V_{DD} and off-chip voltage regulation, wire bonding, or other coupling mechanisms in the package, since the pulse current flows

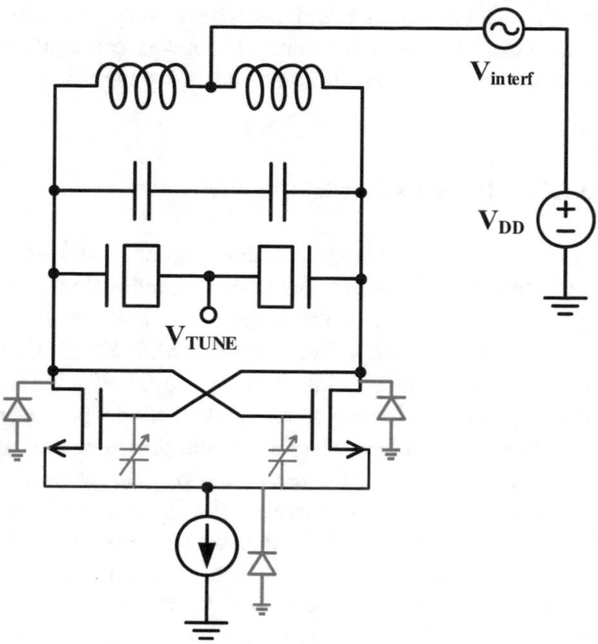

Fig. 3.37 Co-design of
power supply-
stabilized VCO

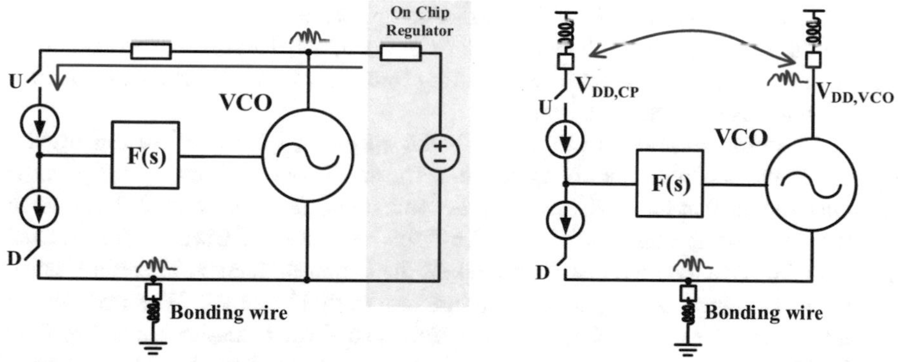

Fig. 3.38 Example of reference spur coupling

through the bonding wire, it may be coupled to the VCO power supply and generate
reference spurs.

3.3.5   Frequency Divider

In an integer-N frequency synthesizer, we usually need a wide division ratio to cover
a wide frequency range. It is impossible to cover the entire range with a simple

counter. Therefore, we did not use a very fast counter to complete the entire operation; we only used several very clever techniques in the integer-N to synthesize various frequency division ratios and very high-frequency division ratios.

3.3.5.1 Pulse Swallowing Divider

A clever method is a programmable frequency divider, as presented in Fig. 3.39. The basic function of the programmable frequency divider is to obtain N input edges and give an output edge at the output. The basic realization of this circuit can be a synchronous counter. In frequency synthesis applications, the problem with standard synchronous counters is that they are too slow for the frequency we want to achieve. The typical solution is to use a dual-mode prescaler in combination with two synchronous counters. The dual-mode prescaler runs at a high speed, whereas the other two counters run at a slower speed. This swallow counter will do the following: For the S cycles in the P cycles, it will actually force the dual-mode prescaler to divide by $(N + 1)$. Therefore, since we have performed this operation for S cycles, the dual-mode prescaler divides by N for the remaining $(P - S)$ cycles. Therefore, the frequency division ratio is $(P \cdot N + S)$. By choosing the right P, S, and N, we can actually cover the number of division ratio specifications we need. The dual-mode prescaler is also called the swallow counter, and its name comes from the idea of "swallowing" 1 from $(N + 1)$ of the dual-mode prescaler.

The dual-mode prescaler N/N + 1 can be 2/3, 4/5, 6/7, 10/11, etc. The advantage of the prescaler is that the average operating frequency of the subsequent synchronous counter is reduced by N times.

The example presented in Fig. 3.40 is the operation scheme of the 2/3 divider circuit. If the modulus control is set to zero, the output of latch 3 will become 1, which means that the AND gate will become transparent. Now, latch 1 and latch 2 are configured as master and slave flip-flops, respectively. Because the inverted signal is fed back to itself, it will obtain half the output frequency. Therefore, when the modulus control is zero, we get the function of dividing by 2. When we set the modulus control to 1, latches 4 and 3 will also form a master–slave flip-flop, participating in the arithmetic circuit. In fact, two master–slave flip-flops can have

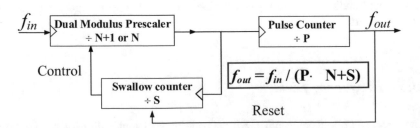

Fig. 3.39 Programmable divider

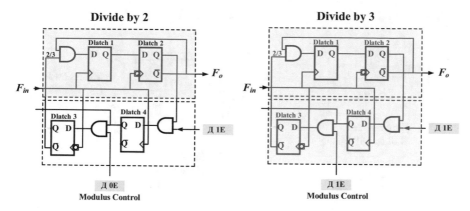

Fig. 3.40 N/N + 1 dual-mode prescaler [13]

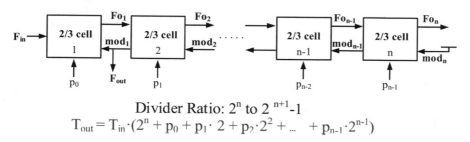

$$\text{Divider Ratio: } 2^n \text{ to } 2^{n+1}-1$$
$$T_{out} = T_{in} \cdot (2^n + p_0 + p_1 \cdot 2 + p_2 \cdot 2^2 + \dots + p_{n-1} \cdot 2^{n-1})$$

Fig. 3.41 Modular programmable divider with cascaded 2/3 cells [13]

four-state circuits, but F_o and the AND logic can ensure that only three states are allowed.

As presented in Fig. 3.41, another divider topology is the so-called truly modular programmable divider, which consists of these 2/3 divider cells [13]. Here, we cascade these cells together and ensure that there is a control path connecting all these cells together. The basic operation is as follows. If we ignore any additional control signal for now, then we see that the input clock will be divided by 2^N. However, if we analyze the modulus operation of this control path p_i ($p_0 \sim p_{n-1}$), for each output signal, the modulus control signal will generate ripples through the frequency divider. For example, every time when this 2/3 cell is reached, the clock will be divided by 2 or 3 according to the value of the modulus control signal p_i. If the p_i is set high, the cell divides its input clock by 3.

Each 2/3 cell must divide its input by at least 2 ($p_0 \sim p_{n-1} = 0$), and then the n 2/3 cells are connected in series to divide input by 2^N. If $p_0 = 1$, the divisor is increased by 1. If the series relationship is $p_1 = 1$, the divisor will be increased by 2, and so on. Therefore, we can actually write the divisor as ($2^N + p_0 + p_1 \cdot 2 + p_2 \cdot 2^2 + \dots + p_{n-1} \cdot 2^{N-1}$). For example, the divisor range of the six-level series is 64–127, the seven-level series is 128–255, and so on.

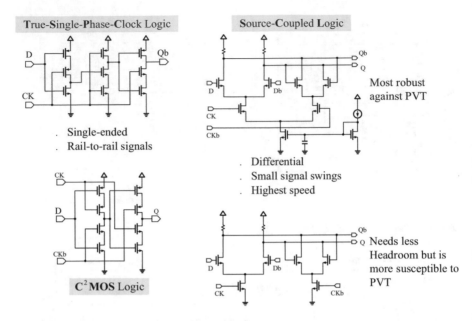

Fig. 3.42 Circuit implementation: high-speed logic

So far, we have discussed frequency dividers on a logical level, and we have several options implemented at the transistor level, as presented in Fig. 3.42. We can choose the true-single-phase-clock (TSPC) logic, which can output rail-to-rail signals, or use the C²MOS logic or more source-coupled logic types with smaller swings for differential operation, usually working at higher speeds. In fact, if we choose this logic, because of the existence of the current source, we must consider the change of the bias current, so that the choice of this latch has the strongest resistance to PVT changes. The TSPC and C²MOS logics do not require a current source, and the required margin is also smaller. To be able to work at the VCO frequency, smaller transistors are used to reduce parasitic capacitance, but they are usually more susceptible to PVT changes.

In the 2/3 cell, we usually need not only the D latch but also the previous AND logic without causing additional delay. Figure 3.43 demonstrates that this logic can be absorbed into the structure in the source-coupled logic, so that this extra gate will not cause extra delay. This means that we can operate the circuit faster.

Figure 3.44 presents the example of a frequency divider [14] with a low-voltage (650-mV) power supply. The frequency at the input of the VCO is 2.4 GHz, and the frequency obtained at the output is 16 MHz. We have a four-stage 2/3-cell cascaded source-coupled logic divider in the front, and the frequency division range is between 140 and 160 MHz. Then, a seven-level conventional CMOS divider logic is adopted. The realization of the 2/3-cell cascaded source-coupled logic divider is presented in Fig. 3.41. The standard logic unit CMOS frequency divider cooperates

Fig. 3.43 Example of source-coupled logic: composite latch

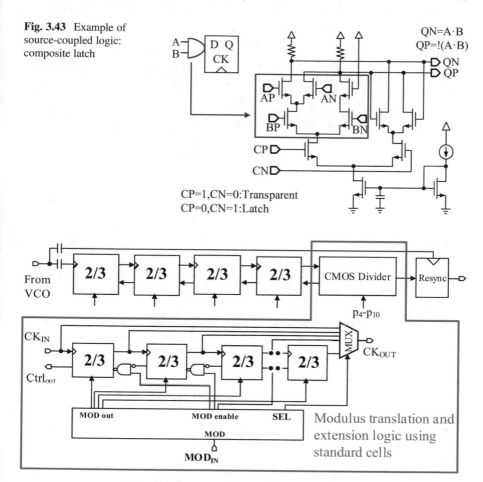

Fig. 3.44 Example of complete frequency divider design [14]

with the module expansion logic to change the length of the 2/3 frequency divider chain and select the corresponding signal output.

3.3.6   Stability Analysis and Design Example of Third-Order Charge-Pump PLL

The stability analysis of the third-order charge-pump PLL is very important, especially when using the integer-N frequency synthesizer in a wide lock range. The loop parameters of the integer-N frequency synthesizer in Fig. 3.7 are defined as follows:

$$K_d = \frac{I_{cp}}{2\pi} \tag{3.25}$$

$$F(s) = \left(R + \frac{1}{sC_1}\right) \Big\| \frac{1}{sC_2} = \frac{sRC_1 + 1}{s^2 RC_1 C_2 + s(C_1 + C_2)}$$

$$= \frac{1}{s(C_1 + C_2)} \frac{1 + s/\omega_z}{1 + s/\omega_p} \tag{3.26}$$

$$\omega_z = \frac{1}{RC_1}, \omega_p = \frac{1}{R\left(\frac{C_1 C_2}{C_1 + C_2}\right)} \tag{3.27}$$

$$H(s) = \frac{1}{N} \tag{3.28}$$

Loop gain:

$$G(s)H(s) = K_d \cdot F(s) \cdot \frac{K_{VCO}}{s} \cdot \frac{1}{N} = \frac{I_{cp}K_{VCO}}{2\pi s^2 N(C_1 + C_2)} \cdot \frac{1 + s/\omega_z}{1 + s/\omega_p} \tag{3.29}$$

$$G(s)|_{s=j\omega} = -\frac{I_{cp}K_{VCO}}{2\pi \omega^2 N(C_1 + C_2)} \cdot \frac{1 + j\omega/\omega_z}{1 + j\omega/\omega_p} \tag{3.30}$$

Thus

$$\angle G(j\omega) = \tan^{-1}(\omega/\omega_z) - \tan^{-1}(\omega/\omega_p) + 180° \tag{3.31}$$

To find the maximum value of $\angle G(j\omega)$

$$\left.\frac{d\phi(\omega)}{d\omega}\right|_{\omega_c} = 0 \rightarrow \omega_c = \sqrt{\omega_z \omega_p} = \frac{1}{RC_1}\sqrt{1 + \frac{C_1}{C_2}} \tag{3.32}$$

The maximum phase margin:

$$\phi_M = \tan^{-1}\left(\frac{\omega_p - \omega_z}{2\sqrt{\omega_p \omega_z}}\right) = \tan^{-1}\left(\frac{b - 1}{2\sqrt{b}}\right) \tag{3.33}$$

If the loop bandwidth ω_c and phase margin ϕ_M are specified, we have:

$$b = \frac{1}{\left(\frac{1}{\cos \phi_M} - \tan \phi_M\right)^2} \tag{3.34}$$

$$\omega_z = \omega_c/\sqrt{b} \tag{3.35}$$

$$\omega_p = \sqrt{b} \cdot \omega_c \tag{3.36}$$

$$f_c = \frac{\omega_c}{2\pi} = \frac{I_{cp}K_{VCO}}{2\pi N} \cdot R \cdot \frac{b-1}{b} = \frac{I_{cp}K_{VCO}}{2\pi N} \cdot R \cdot \frac{C_1}{C_1 + C_2} \tag{3.37}$$

$$R = \frac{N\omega_c}{I_{cp}K_{VCO}} \frac{b}{b-1} \tag{3.38}$$

$$C_1 = \frac{1}{R_1 \omega_z}, C_2 = \frac{1}{R_1 (\omega_p - \omega_z)} \tag{3.39}$$

3.3.6.1   Design Example

Design Specification

Build up the behavior model of a third-order charge-pump PLL if the reference frequency is 20 MHz, the division ratio of the feedback divider is 64, and VCO gain is 200 MHz/V.

(a) Implement a second-order loop filter and find the charge-pump current such that the loop bandwidth = 100 kHz and the PLL phase margin = 45°. Use Matlab to verify the design.
(b) Following (a), plot the closed-loop transfer function.
(c) Following (b), plot the noise transfer functions of reference input, VCO, and loop filter, respectively, using Matlab. What are the natural frequency and damping factors in the design?
(d) Follow (a), if the frequency division ratio changes from 64 to 60, find the variations of natural frequency and damping factor of the PLL. Describe how to maintain the loop parameters.
(e) If the reference frequency hops from 20 MHz to 24 MHz at t = 0, what is the settling time of the PLL when the output frequency error is allowed for 20 ppm?

Design Solution

(a) According to the phase margin requirement of 45°, refer to the values of $b = 5.82$ and $\sqrt{b} = 2.41$ in the loop filter design in Table 3.1.
Let $I_{cp} = 1$ mA, apply (3.38) to obtain R_1.

Table 3.1   Lookup table for phase margin

ϕ_M	10°	20°	30°	40°	45°	50°	55°	60°	70°	80°
$\sqrt{b}$	1.19	1.43	1.73	2.14	2.41	2.75	3.17	3.73	5.67	11.43
b	1.42	2.04	3.00	4.60	5.83	7.55	10.06	13.93	32.16	130.65

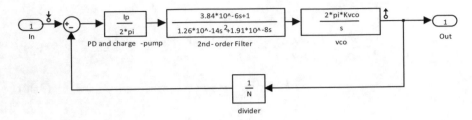

Fig. 3.45 Behavioral model of the Type-2 charge-pump PLL

$$R_1 = \frac{64 \cdot 2\pi \cdot 100 \text{ kHz}}{1\text{mA} \cdot 200\text{MHz/V}} \cdot \frac{5.82}{5.82 - 1} = 243\Omega$$

By applying (3.35) and (3.36) into (3.39), we can obtain C_1 and C_2.

$$C_1 = \frac{1}{243\Omega \cdot 2\pi \cdot 100 \text{ kHz}/2.41} = 15.8 \text{ nF}$$

$$C_2 = \frac{1}{243\Omega(2\pi \cdot 2.41 \cdot 100 \text{ kHz} - 2\pi \cdot 100 \text{ kHz}/2.41)} = 3.29 \text{ nF}$$

The behavior model is built in Simulink as follows (Fig. 3.45):

The open-loop frequency response is shown in the Fig. 3.46. The phase margin is $45°$ and the loop bandwidth is 100 KHz ($= 2\pi \times 100 \text{ K} = 6.28 \times 10^5$).

(b) The closed-loop transfer function of the Type-2 charge-pump PLL is shown in Fig. 3.47.

Because it is a third-order charge-pump PLL, the transfer function is more complicated than the second-order function. However, we can approximate it to the second order because C_2 is smaller than C_1. Therefore, the natural frequency and damping factor

$$\omega_n = \sqrt{\frac{I_{cp}(2\pi K_{VCO})}{2\pi N C_1}} = 444742 [\text{rad/s}]$$

$$\zeta = \frac{R}{2} \sqrt{\frac{I_{cp}(2\pi K_{VCO}) C_1}{2\pi N}} = 0.853$$

(c) The noise transfer functions from the reference input (In), VCO, and loop filter to the output (Out) are shown in Fig. 3.48.

(d) Following (a), if the frequency division ratio changes from 64 to 60,

$$\omega_n = \sqrt{\frac{I_{cp}(2\pi K_{VCO})}{2\pi N C_1}} = 459328.8 [\text{rad/s}]$$

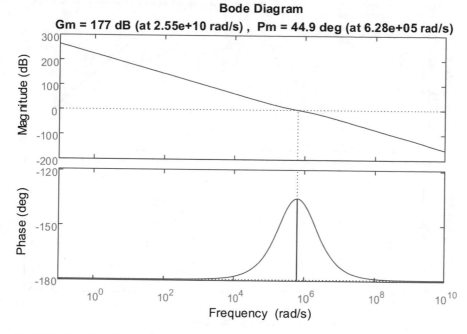

Fig. 3.46 Open-loop frequency response of the Type-2 charge-pump PLL

$$\zeta = \frac{R}{2}\sqrt{\frac{I_{cp}(2\pi K_{VCO})C_1}{2\pi N}} = 0.881$$

$$I_{cp1} = I_{cp}\frac{60}{64} = 0.935 \text{ mA}$$

As the division ratio becomes smaller, the natural frequency and the damping coefficient increase.

We can reduce I_{cp} to keep the same loop parameters, which means that increasing I_{cp} from 1 mA to 0.94 mA can obtain the same loop bandwidth and damping coefficient.

(e) Under this hop condition, the frequency change Δf_{step} is (24 MHz - 20 MHz) $\times$ 64 = 256 MHz. The frequency error f_ε within 20 ppm is 256 MHz $\times$ $\pm$20 ppm = $\pm$5.12 kHz.

According to Eq. (3.7), we can obtain t_s:

$$t_s = \frac{-1}{\zeta\omega_n}\ln\left(\frac{\zeta f_\varepsilon}{\Delta f_{step}}\right) = 24.7\mu s$$

The transient simulation result of the PLL with the input frequency jumping from 20 MHz to 24 MHz and the frequency error of 20 ppm is shown in Fig. 3.49.

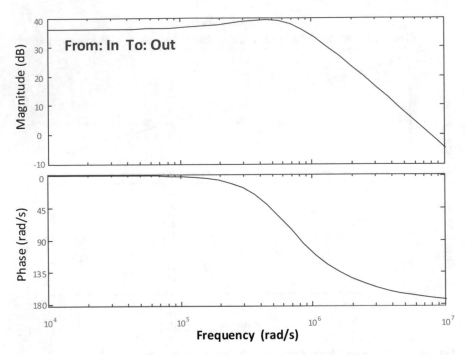

Fig. 3.47 Closed-loop frequency response of the Type-2 charge-pump PLL

Table 3.2 Specifications of type-2 charge-pump PLL		Specifications
	Reference frequency	20 MHz
	Divide-ratio	64
	VCO gain	200 MHz/V
	Loop bandwidth	100 kHz
	Phase margin	45

The Matlab code is shown in Appendix A.

3.4   Fractional-N Frequency Synthesizer

For the wireless example, in order to have a certain channel spacing between the carriers, the frequency synthesizer needs to achieve a very good frequency resolution. Obtaining a good frequency resolution at the output will make $N \cdot f_{ref}$ smaller. If f_{ref} becomes smaller, this means that the filter must be set to a lower bandwidth. Therefore, the classic integer-N requires a trade-off between resolution and overall PLL bandwidth. Generally speaking, if we want a good frequency resolution, we must have a low bandwidth. A low bandwidth makes the setup time worse. Using fractional-N PLLs can solve the basic resolution/bandwidth/noise trade-offs. The reason is that the fractional-N PLL can provide higher loop bandwidth while

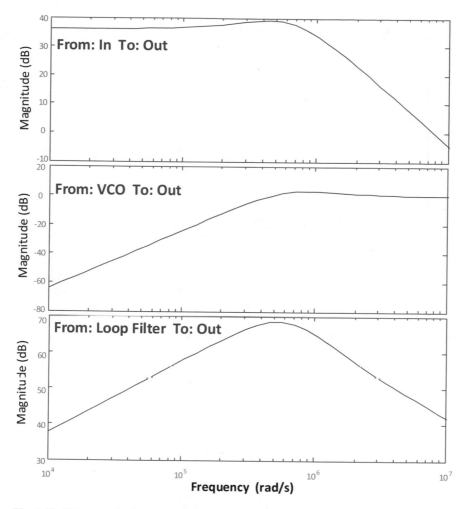

Fig. 3.48 Noise transfer functions from reference input, VCO, and Loop filter to output

achieving very good frequency resolution, as presented in Fig. 3.50. Therefore, the idea of fractional-N PLL is oversampling by employing periodic modulation to the modulus divider. Suppose that we have a 2-bit accumulator, which provides a carry output every four times: 0, 0, 0, 1. Let "0" control N and "1" control N + 1, so this output will be N, N, N, N + 1, N, N, N, N + 1, and so on. PLL cannot generate frequency offset; thus, to use this modulation, PLL must find the average frequency information defined by N + 1/4. Therefore, it can create a frequency division ratio between N and N + 1, similar to the concept of sampling. However, this periodic modulation is problematic as it can modulate the VCO periodically, which is why we need a ΔΣ modulator. The ΔΣ modulator is used as a random number generator. Therefore, this ΔΣ modulator can break the periodicity of the carry output, and at the same time, we can control the weighting ratio of 0 and 1. Therefore, by changing the

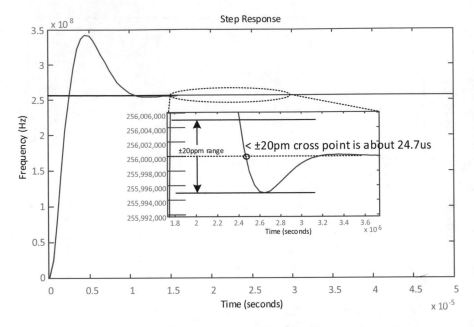

Fig. 3.49 PLL transient response hopping to 256 MHz with the frequency error of 20 ppm

Fig. 3.50 Fractional-N PLL

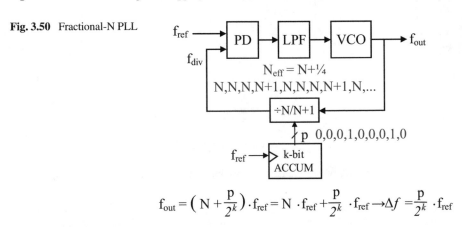

$$f_{out} = \left(N + \frac{p}{2^k} \right) \cdot f_{ref} = N \cdot f_{ref} + \frac{p}{2^k} \cdot f_{ref} \rightarrow \Delta f = \frac{p}{2^k} \cdot f_{ref}$$

number of digital inputs, we can generate an arbitrary weighted ratio of 0 and 1 so that we can accurately generate the output frequency.

To summarize the $\Delta\Sigma$ function, first, we perform digital interpolation (interpolation), so no matter what the output frequency of the VCO, we can achieve a very fine resolution. Furthermore, we can achieve 1 or 0.1 Hz resolution by simply increasing the number of bits in the accumulator. Then, $\Delta\Sigma$ provides randomization and noise shaping. Due to the randomness, we cannot see any close common-mode noise because noise shaping pushes the close phase noise to higher frequencies.

3.5 Direct Digital Synthesis (DDS)

Figure 3.51 presents another frequency synthesizer, which is the direct digital synthesis (DDS) [15]. It uses a feedforward method and does not involve any feedback. Furthermore, it has no error correction. By using the DDS, three aspects, namely, amplitude, phase, and frequency, can always be controlled digitally. Since the frequency can be digitally controlled, the frequency can be instantly changed in picoseconds or nanoseconds within the switching characteristics of the device. Unfortunately, the DDS is not actually a digital system because it uses a DAC and LPF. Moreover, it utilizes the frequency control word (FCW) to control the input address sequence of the ROM, because the ROM has stored the data of the waveform amplitude. The order of the input addresses determines the speed of the output waveform. Since the output amplitude is digital data, it needs to be converted to an analog amplitude through a DAC. Thus, it puts the zero-order hold output and then filters it out. For wireless applications, it is currently not practical because it requires the clock frequency to be more than three times the output frequency. For example, if we want to synthesize 2.4 GHz, we need to run at 7.2 GHz. Now, the operating frequency of digital circuits is a bit too fast, but it can be used as a hybrid. For example, the DDS can change the frequency very quickly and perform the modulation. There is no feedback loop, so it is currently in a fast state. Therefore, if we are limited to a very low frequency (e.g., tens of MHz), we can use it as an input reference to the frequency synthesizer. The output frequency is a multiple of the DDS frequency. This has actually been used in base stations for some time. It also has good properties, such as fast settling time.

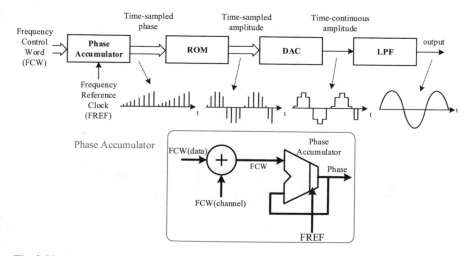

Fig. 3.51 Direct digital synthesis (DDS)

3.6   All-Digital Phase-Locked Loop

For the development of the ADPLL, we can replace the analog module with the corresponding digital module, as presented in Fig. 3.52. Digitization is relatively easy. In a digital PLL, the PFD must be composed of a so-called time-to-digital converter (TDC). The digitizer is actually an analog circuit. Therefore, it is still beneficial and needed to solve the variability problem. The analog loop filter is replaced with a digital loop filter. Digital filters are very compact and easy to program, so we no longer need to worry about noise or any form of leakage. Finally, the VCO in a digital PLL is called a DCO, which is a numerically controlled oscillator. We will have a digital word to control the frequency instead of using a continuous analog voltage. These DCOs still have phase-noise problems and tuning problems; the quantization effect is a problem that did not exist before, so attention should be paid to it. While digital filters are discussed in many textbooks, this book only introduces DCO and TDC.

3.6.1   Frequency Switching in DCO

If the original design has implemented a HP analog PLL with an LC structure and uses a varactor to change the frequency of the resonant LC, then a simple way to implement a numerically controlled oscillator is to let the DAC drive the actual varactor input, as presented in Fig. 3.53(a). The advantage of doing this is that if we spend a lot of time developing a VCO with a high performance, the VCO can still be used directly, fundamentally having the advantage of an ADPLL.

We can basically use segmentation to simplify the overall control of the system, as presented in Fig. 3.53(b). For the PLL, we can compensate for the process changes in the system. For example, there is only one coarse control array, which is ultimately binary in terms of overall implementation. The fine-tuning control array

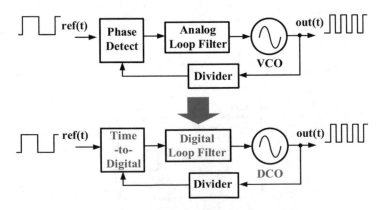

Fig. 3.52 Correspondence between analog PLL and all-digital PLL [16]

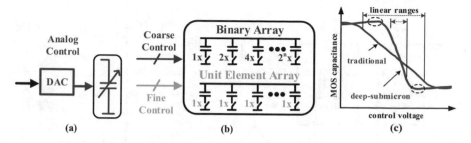

Fig. 3.53 Different types of varactors for DCO. (**a**) Simple method, (**b**) switched capacitor, (**c**) MOS capacitor

uses unit capacitance for small-range adjustments. The advantage of binary is that through a very simple control scheme, a large range can be easily obtained from the binary code. Its disadvantage is that if there is a mismatch between the binary-weighted capacitance values, it may eventually have non-monotonic characteristics.

The frequency tuning in the traditional VCO uses varactor as it allows us to obtain a very wide linear capacitance. Therefore, precise and wide frequency tuning can be achieved. In the deep-submicron CMOS process, this linear range is greatly com-pressed, and the linear range continues to deteriorate. We only use two flat and stable working areas, such as the CV curve presented in Fig. 3.48(c). At present, in nanoscale processes, MOS capacitors only operate in the low-capacitance and high-capacitance modes. Therefore, another advantage is that if there is noise in the control voltage, the noise will be converted into a very small capacitance error.

3.6.2   Time-to-Digital Converter

The time-to-digital converter is to compare the two input signals to obtain the digital time difference. Basically, the reference clock signal and the frequency divider output signal are obtained and sent through a series of delay stages, and then, these various stages are sampled according to the reference frequency edge. There-fore, in the example presented in Fig. 3.54, by using the reference clock for sampling, the various delay values correspond to a thermometer-coded version of the phase error.

Since various delay elements actually have mismatch problems in the delay of each stage, the mismatch will actually cause the overall phase error characteristic to appear nonlinear. As can be seen from Fig. 3.55, if the delay of each delay element is different, a nonlinear conversion curve will be generated. Suppose the PLL wants to lock to a certain phase error, for some phase error PLL cannot feel it, but for some phase error PLL feels that the phase changes greatly. Thus, the system will become nonlinear and the phase will jump back and forth, resulting in phase lock failure. Therefore, in terms of the overall TDC design, there are basically two problems. The

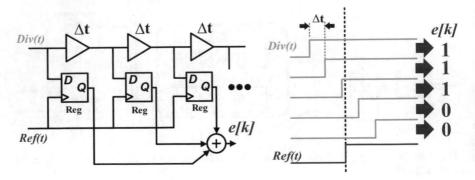

Fig. 3.54 Classic time-to-digital converter

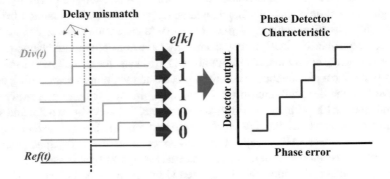

Fig. 3.55 Effect of finite resolution and delay mismatch

first is the resolution of the phase error, and the second is the linearity of the phase-error detector.

To obtain better TDC resolution, one applicable technique is the Vernier delay, shown in Fig. 3.56. In this technique, the transmitted signal passes through a series of buffers to delay the reference edge, and the delay of the bottom buffer different from the delay of the top buffer is selected. By doing this, we can achieve a finer resolution, because a finer resolution actually corresponds to the difference between these two delays. However, we will still have mismatch problems on the two chains, and the number of mismatches will increase because we are actually using a shorter time base for comparison. Thus, the same mismatch may be larger than before, the impact become greater. In addition, if we want to obtain a reasonable range to check the phase error, since the increment of each step is small, we will have a very long Vernier structure. A structure with a longer buffer delay requires a lot of power and area.

Another method with the same TDC resolution is to employ the so-called two-step method, presented in Fig. 3.57 [17]. In this method, the standard delay chain is used for rough quantification; then, once the actual time error range is determined, the appropriate delay output will be sent to the second stage

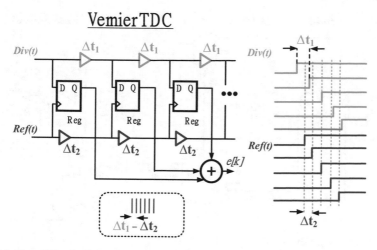

Fig. 3.56 Use of Vernier delay technology to improve resolution [18]

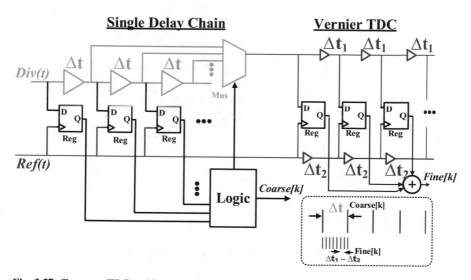

Fig. 3.57 Two-step TDC architecture can reduce the area [18]

corresponding to the Vernier structure, which will provide a better resolution when measuring the time difference. Any two-step structure needs to consider that the boundary of the standard delay chain is basically aligned with the entire range of the Vernier, so a certain amount of calibration may be required.

Another possible method with finer TDC resolution is to replace the Vernier structure with a time amplifier and the same type of single delay chain structure, as presented in Fig. 3.58 [19]. We amplify the time before entering the second stage. Through amplification, a higher resolution can be obtained. The metastable state of

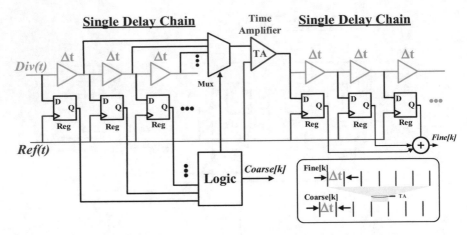

Fig. 3.58 Two-step TDC using time amplification [18]

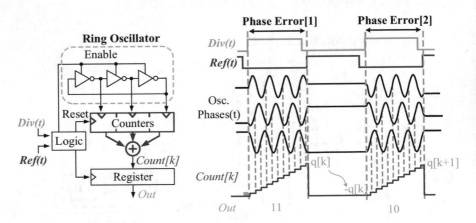

Fig. 3.59 Oscillator-based TDC

the latch is used to create a time amplifier, but there are still some problems in correcting the mismatch.

Another possible approach, in addition to delaying one sideline and then comparing it with the other sideline, is to introduce a separate time base to perform the measurements. Therefore, for example, a ring oscillator will generate a higher frequency waveform. Then we calculate the number of edges that the oscillator output appears within the actual phase error window. The value on the counter represents the quantized phase error, as presented in Fig. 3.59 [20].

3.7 Measurement

This section is to explain how to use a spectrum analyzer to measure quantized phase noise. Figure 3.60 is a simplified block diagram of the spectrum analyzer for phase noise measurement. First, when the input signal comes in, the internal local oscillator will scan through frequencies and generate corresponding intermediate frequencies. With each intermediate frequency through a band-pass filter, the bandwidth of the band-pass filter defines the noise bandwidth and is called RBW or resolution bandwidth. We can sweep the entire SPAN range with the RBW to plot the power of each scanned frequency. Then, we use a smoothing filter or the video bandwidth, defined by VBW, to measure the power at a given RBW and plot the output spectrum. VBW is just for a visual effect, not important. However, RBW is a very critical parameter because it defines the noise bandwidth. But in the spectrum analyzer we cannot distinguish any AM noise because AM noise can also be displayed here. Thus, if we want to accurately calculate the phase noise, we still need to buy an expensive phase noise analyzer. The phase noise analyzer uses a frequency extractor with a delay line, so any AM noise can be suppressed, and PM noise and noise power can be accurately measured.

Let us look at an example of measurement. According to the spectrum in Fig. 3.61, we can calculate the phase noise when the frequency offset is 10 MHz.

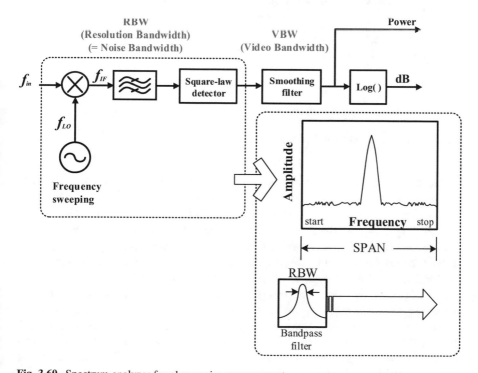

Fig. 3.60 Spectrum analyzer for phase noise measurement

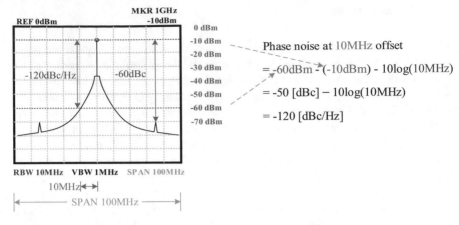

Fig. 3.61 Example of phase noise measurement by a spectrum analyzer

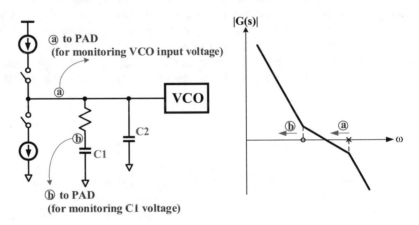

Fig. 3.62 Monitor VCO control voltage

The span is 100 MHz, each grid represents a 10 MHz offset, and the top is a 0 dBm reference, so the noise power at 10 MHz offset is −60 dBm. We can subtract the reference carrier power of −10 dBm and obtain actually −50 dBc, but this is not a normalized phase noise dBc/Hz. We only need to subtract the noise bandwidth value to obtain −120 dBc/Hz phase noise. Similarly, the spur we can measure is −60 dBc. Even if the resolution bandwidth is changed, the spur level will not change because the spurious level is not random noise, but deterministic noise.

For IC design, we often place test pads to monitor the control voltage of the VCO for debugging, or sometimes we want to characterize the tuning capability of the VCO. In either case, we usually place the pad connection at point ⓐ in Fig. 3.62 because this is the input of the VCO. However, this will increase the total capacitance of C2. If the capacitance is increased, the pole frequency will be decreased and close to the PLL bandwidth, which will reduce the phase margin and cause stability problems. If we put the pad at ⓑ, C1 is a large capacitance, so it will not have much impact on the loop dynamics. Second, even if this adds a lot of capacitance to C1, an increase in C1 will reduce the zero frequency and further improve stability. More importantly, because ⓑ is low impedance and ⓐ is high impedance, we always set the connection to low-impedance nodes to better resist noise coupling problems.

References

1. Best, R. E. (1993). *Phase-locked loops: Theory, design, and applications*. McGraw-Hill Companies.
2. Wilson, W. B., et al. (2000). A CMOS self-calibrating frequency synthesizer. *IEEE Journal of Solid-State Circuits, 35*(10), 1437–1444.
3. Juarez-Hernandez, E., & Diaz-Sanchez, A. (2001). A novel CMOS charge-pump circuit with positive feedback for PLL applications. In *ICECS 2001. 8th IEEE International Conference on Electronics, Circuits and Systems (Cat. No. 01EX483)* vol. 1. pp. 349–352.
4. Rhee, W. (1999, May). Design of high-performance CMOS charge pumps in phase-locked loops. In *1999 IEEE International Symposium on Circuits and Systems (ISCAS)*, vol. 2, pp. 545–548.
5. Gierkink, S. L. J. (2008). Low-spur, low-phase-noise clock multiplier based on a combination of PLL and recirculating DLL with dual-pulse ring oscillator and self-correcting charge pump. *IEEE Journal of Solid-State Circuits, 43*(12), 2967–2976.
6. Hung, C., & Liu, S. (2009). A leakage-suppression technique for phase-locked systems in 65nm CMOS technology. In *ISSCC 2009*, pp. 400–401.
7. Koo, Y., Huh, H., Cho, Y., Lee, J., Park, J., Lee, K., Jeong, D. K., & Kim, W. (2002). A fully integrated CMOS frequency synthesizer with charge-averaging charge pump and dual-path loop filter for PCS-and cellular-CDMA wireless systems. *IEEE Journal of Solid-State Circuits, 37*(5), 536–542.
8. Lakshmikumar, K. R. (2009). Analog PLL design with ring oscillators at low-gigahertz frequencies in nanometer CMOS: Challenges and solutions. *IEEE Transactions on Circuits and Systems II: Express Briefs, 56*(5), 389–393.
9. Andreani, P., & Mattisson, S. (2000). On the use of MOS varactors in RF VCOs. *IEEE Journal of Solid-State Circuits, 35*(6), 905–910.
10. Kral, A., Behbahani, F., & Abidi, A. A. (1998). RF-CMOS oscillators with switched tuning. In *Proceedings of the IEEE 1998 Custom Integrated Circuits Conference (Cat. No. 98CH36143)*, pp. 555–558.
11. Yu, S. A., & Kinget, P. (2008, September). A 0.042-mm 2 fully integrated analog PLL with stacked capacitor-inductor in 45nm CMOS. In *ESSCIRC 2008-34th European Solid-State Circuits Conference*, pp. 94–97.
12. Zhang, F., & Kinget, P. R. (2006). Design of components and circuits underneath integrated inductors. *IEEE Journal of Solid-State Circuits, 41*(10), 2265–2271.
13. Vaucher, C. S., Ferencic, I., Locher, M., Sedvallson, S., Voegeli, U., & Wang, Z. (2000). A family of low-power truly modular programmable dividers in standard 0.35-µm CMOS technology. *IEEE Journal of Solid-State Circuits, 35*(7), 1039–1045.
14. Yu, S. A., & Kinget, P. (2007). A 0.65 V 2.5 GHz fractional-N frequency synthesizer in 90nm CMOS. In *2007 IEEE International Solid-State Circuits Conference. Digest of Technical Papers*, pp. 304–604.
15. Tierney, J., Rader, C., & Gold, B. (1971). A digital frequency synthesizer. *IEEE Transactions on Audio and Electroacoustics, 19*(1), 48–57.
16. Staszewski, R. B., Leipold, D., Muhammad, K., & Balsara, P. T. (2003). Digitally controlled oscillator (DCO)-based architecture for RF frequency synthesis in a deep-submicrometer

CMOS process. *IEEE Transactions on Circuits and Systems II: Analog and Digital Signal Processing, 50*(11), 815–828.

17. Ramakrishnan, V., & Balsara, P. T. (2006). A wide-range, high-resolution, compact, CMOS time to digital converter. In *19th International Conference on VLSI Design held jointly with 5th International Conference on Embedded Systems Design (VLSID'06)*, pp. 6–12.

18. Perrott, M. H. (2008). Digital phase-locked loops. In *ISSCC Tutoial*.

19. Lee, M., & Abidi, A. A. (2007). A 9b, 1.25 ps resolution coarse-fine time-to-digital converter in 90nm CMOS that amplifies a time residue. In *2007 IEEE Symposium on VLSI Circuits*, pp. 168–169.

20. Straayer, M. Z., & Perrott, M. H. (2009). A multi-path gated ring oscillator TDC with first-order noise shaping. *IEEE Journal of Solid-State Circuits, 44*(4), 1089–1098.

Chapter 4
A 0.35-V 240-µW Fast-Lock and Low-Phase-Noise Frequency Synthesizer for Implantable Biomedical Applications

For implantable frequency synthesizers, realizing ultra-low voltage (ULV) and low power in addition to meeting PLL targets, fast lock and low phase noise, poses a difficult challenge. This chapter will introduce techniques to achieve PLL targets as well as ULV and low power in the same chip through the use of a regular CMOS technology node. A curvature-PFD technique achieves both faster locking and lower jitter compared with conventional techniques. A two-step switching technique substantially reduces the power consumption in current mirrors and reduce noise when switching from a charge-pump. Leakage analysis and subthreshold-leakage-reduction technique reduce reference spur and jitter to the voltage-controlled oscillator (VCO). A dither technique randomizes and averages reference spurs. The proposed chip was implemented in 90-nm CMOS technology; the 0.35-V medical-band frequency synthesizer consumes 238-µW power while generating output clock of 401.8–431.31 MHz and exhibiting a phase noise of -105.7 dBc/Hz at 1-MHz frequency offset with 20-µs locking time.

4.1 Introduction

Although the productivity and stability of the society depend on the outcome of healthcare, rising healthcare costs are affecting almost everyone in the world. In recent years, research into the use of engineering innovations to improve efficacy while reducing the cost of health services has grown significantly. In these efforts, wireless sensing and communication through human tissues to transmit physiological or biochemical information have provided a basis for disease diagnosis. Due to compelling medical principles, the prospect of direct electronic communication with the brain now exists. The patient safety, comfort, and mobility can be improved by using radio frequency (RF) in implantable biomedical devices, which not only enable patients to live a routine life at home and work, but also improve diagnostics. Wireless neural interface devices will solve many complex issues related to wired connections between components consisting of neural interface systems and are therefore important for developing devices that can be implanted for long periods of time. To cover more frequency range for medical using, the Federal Communications Commission (FCC) announced a new medical device radio communication service (MedRadio) in 2009, which can use implanted biomedical devices to transmit data to support diagnostic or therapeutic purposes. Also, it will increase the efficiency of hospital management and reduce medical costs.

In wireless communications, the spectrum is a precious asset because more and more wireless users need to make more efficient use of already scarce frequency resources. Communication transceivers rely heavily on frequency translation using a local oscillator (LO). Accordingly, the spectral purity of the oscillators in the transmitter and receiver is one of the factors restricting the maximum number of channels and users available. Therefore, the frequency synthesizer must be designed to reduce the noise problem on the spectrum.

Battery life is an important challenge for all wireless implantable biomedical devices and must last a long time to be replaced or recharged. Therefore, reducing power consumption extends battery life or battery-less remote power operation. In the design of wireless transceivers, the frequency synthesizer that produces the carrier frequency for wireless transmission is one of the modules with the largest power consumption, so diminishing its power consumption is crucial for ultra-low power transceivers. And if the implanted system generates excessive heat dissipation, it will cause damage to surrounding tissues, so ultra-low power consumption is necessary for the implanted system. Due to the dynamic power consumption is proportional to the square of the power supply voltage (V_{DD}), decreasing V_{DD} is one of the most practical ways to lower power consumption in digital circuits. Low supply voltages and comparatively high threshold voltages not only restrict the number of stackable transistors, but also worsen under process variation. The design of charge-pumps (CP), voltage dividers, and voltage-controlled oscillators (VCOs) becomes very challenging. As a result, PLLs design is crucial in low-voltage, low-power implantable SoC.

There are some innovations we applied to implantable frequency synthesizers. A curvature-PFD technique achieves both faster locking and lower jitter compared with conventional techniques. A two-step switching technique substantially reduces the power consumption in current mirrors and reduces noise when switching from a charge-pump. Leakage analysis and subthreshold-leakage-reduction technique reduce reference spur and jitter to the VCO. A dither technique randomizes and averages reference spurs.

4.2 Proposed Design Techniques

The block diagram of the proposed 0.35-V low-power frequency synthesizer is presented in Fig. 4.1, which includes a CPFD (Curvature-PFD) with dither function, charge-pump with leakage reduction, off-chip loop filter, ring-type VCO, and frequency divider with a divider ratio of 218–235 for MedRadio bands as shown in Table 4.1.

In this chapter, CPFD, two-steps switching, leakage reduction, dither techniques, and low-phase noise VCO are proposed to achieve the specifications while working at ULV for the implantable integer-N frequency synthesizer with a reference frequency of 1.843 MHz. The details of each technique are explained in the following sections.

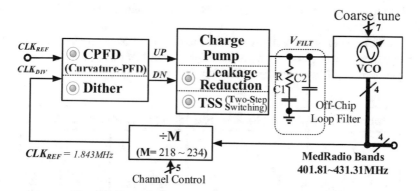

Fig. 4.1 Block diagram of the proposed frequency synthesizer

Table 4.1 Available channels for MICS band

Channel control bits							Divide number	f_out	
							$8*(27-i)+9*$		MICS band
in4	in3	in2	in1	in0	Bin	Dec	$i+1$	MHz	(MHz)
0	0	0	0	1	00001	1	218	401.82	401–402
0	0	0	1	1	00011	3	220	405.50	405–406
0	1	0	0	0	01000	8	225	414.72	413–419
0	1	0	0	1	01001	9	226	416.56	
0	1	0	1	0	01010	10	227	418.41	
0	1	1	1	1	01111	15	232	427.62	426–432
1	0	0	0	0	10000	16	233	429.47	
1	0	0	0	1	10001	17	234	431.31	

4.2.1 Curvature-PFD (CPFD) Technique

For the PLL-based frequency synthesizer, a trade-off occurs between locking speed
and output jitter. To minimize the output jitter due to external supply noise, the loop
bandwidth should be as narrow as possible. However, to minimize output jitter due
to internal oscillator noise, or to obtain improved tracking and acquisition properties,
the loop bandwidth should be as wide as possible [1]. The general guideline for PLL
designs is to set the loop bandwidth to approximately one-tenth of the reference
frequency to ensure stability and it is also to reduce the reference frequency spur
[2]. The fixed loop bandwidth also fixes the locking speed and output jitter. If we can
adaptively control the loop bandwidth according to the locking status and PFD phase
error, we can acquire both benefits to achieve fast locking while also maintaining
improved jitter performance in the same PLL. When the phase error is large, such as
in fast-lock mode, the PLL increases the loop bandwidth and achieves fast locking.
Conversely, when the phase error is small, the PLL decreases the loop bandwidth
and minimizes output jitters. The proposed CPFD scheme presented in Fig. 4.2
assists the precise charge-pump-current control based on the phase difference of the
PFD output to achieve both faster locking and lower phase noise. The smooth
transfer curve also guarantees that the low noise charge and discharge current
helps to decrease jitter.

  The CPFD locking control is based on the condition of the lock detector; when
the phase error θ_e (the pulse width difference between UPi and DNi) is determined to
be larger than $|\theta_M|$ (the midpoint of the full detection range), i.e., $\theta_e \geq |\theta_M|$, and the

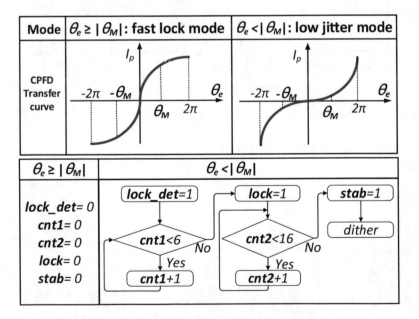

Fig. 4.2 CPFD transfer curve of two operation mode and its tracking control flowchart

lock_det = 0, fast-lock mode is enabled, and the charge-pump (CP) provides a greater current to reduce locking time, resulting in the PLL being prompted to jump into the $\theta_e < |\theta_M|$ mode, where the tracking speed will be drastically reduced. Both *cnt1* and *cnt2* are monitoring counters for counting the numbers of *lock_det* = 1 and *lock* = 1, respectively. The phase detector then continues to monitor its status if $\theta_e < |\theta_M|$ (*lock_det* = 1) is checked more than six times (*cnt1* ≥ 6). The choice of six is trade-off between control stability and locking speed. If the criteria is set too small, it is possible for PFD to switch back and forth between fast-lock and low-jitter modes, causing the system not to lock stably. However, if the criteria is set too large, it will take longer locking time. Then, by setting *lock* = 1, the PLL becomes stable on the basis of $\theta_e < |\theta_M|$, and that the transfer curve changes to a low-jitter mode. In this mode, the transfer curve is similar to the aforementioned decreased loop bandwidth. This smooth curve, corresponding to a smooth charge or discharge current into the loop filter, will not generate a current spike to destroy the VCO phase noise. The digital-assistance circuit regularly checks to see if this status is maintained (*lock* = 1) for more than 16 times (*cnt2* ≥ 16). The choice of 16 is to guarantee a stably locked status. If the criteria is set too small, the added random dither may combine with dead-zone pulse, causing a transient frequency shift (or jitter) during the locking procedure. If the outcome triggers $\theta_e \geq |\theta_M|$, *cnt1* and *cnt2* will be both reset, resulting in a new tracking procedure. If *cnt2* > 16, then the *stab* is set to one (*stab* = 1), meaning that the PLL is well locked and can begin a dither operation to randomize the reference spurs. If the user prefers lower phase noise to lower frequency spurs, an option is not to enable the dither. The ideal locking procedure is to keep stable mode-swap from one step to the other.

Compared with the aforementioned studies, CPFD possesses two notable features: First, robustness because of digital control procedure without being affected by PVT variation. Second, low noise current transfer curve because smooth charge or discharge current on the VCO control line can avoid unnecessary jitter noise.

The block diagram and operation modes in Fig. 4.3 illustrated where the *lock_det* and *lock* signals control the status of CPFD and charge current to loop the filter between the two operating modes. The reference clock and divided VCO clock feed into the PFD generating two signals: *en* (*en* = $UP_i \oplus DN_i$) and *rst* (*rst* = $UP_i -$ DN_i). The *en* signal represents the "pure" phase difference between the two input clocks, and the *rst* signal denotes the reset function. An additional signal *ckd* is a delayed version of the *en* signal for phase-detector alignment purposes. The *ckd* was fed through a series of tapped delay chains, and the total number of delay stages can cover the period of the VCO output clock. Because the tapped delay chain decreases the *ckd* period stage by stage, the period shortens from the first stage, ckd_0, to the final stage, ckd_{15}. The lock detector operates through the *rst* signal to sample the middle stage of the delay chain, ckd_7. A high signal indicates that the phase difference between the reference clock and feedback clock is significant, and the maintenance of a low *lock_det* signal indicates the status of fast-lock mode ($\theta_e \geq |\theta_M|$). The unit current source is 1.5 μA, the current profile is designed as a smooth transfer curve, and the total charge current I_p shown in Fig. 4.4 can be expressed as:

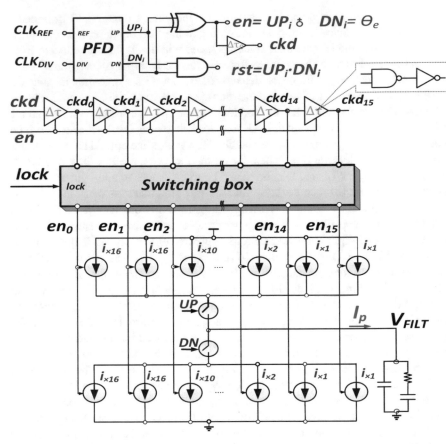

Fig. 4.3 *lock_det* and *lock* signals control the CPFD status and charge current to the loop filter

$$I_p = en_0 \cdot i_{\times 16} + en_1 \cdot i_{\times 16} + en_2 \cdot i_{\times 10} + en_3 \cdot i_{\times 10} +$$

$$en_4 \cdot i_{\times 8} + en_5 \cdot i_{\times 8} + en_6 \cdot i_{\times 8} + en_7 \cdot i_{\times 8} +$$

$$en_8 \cdot i_{\times 8} + en_9 \cdot i_{\times 8} + en_{10} \cdot i_{\times 4} + en_{11} \cdot i_{\times 4} +$$

$$en_{12} \cdot i_{\times 2} + en_{13} \cdot i_{\times 2} + +en_{14} \cdot i_{\times 1} + en_{15} \cdot i_{\times 1} \qquad (4.1)$$

This adjustable current profile based on locking status is the key to achieve better performance than conventional fixed charge current. However, higher phase margin should be involved in the design phase because of wider loop dynamic.

The *lock* signal controls the switching box and alters the charge current sequentially from en_0 to en_{15} according to (4.1). It results in the sequential turn-on current as: $\times 16 \rightarrow \times 32 \rightarrow \times 42...$, so that the CP provides a large and rapidly increasing current for fast lock. However, if the two clocks are within close enough proximity, the *ckd* signal goes low before reaching the middle of the delay chain, resulting in a low signal (*lock_det* = 1). If *lock_det* = 1 for more than six clock cycles, the *lock*

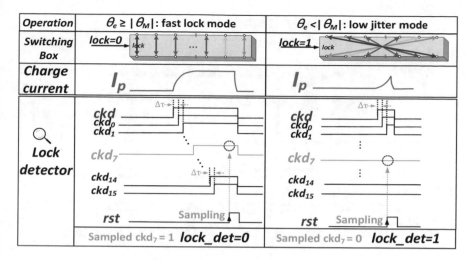

Operation	$\theta_e \geq |\theta_M|$: fast lock mode	$\theta_e < |\theta_M|$: low jitter mode
Switching Box	lock=0	lock=1
Charge current	I_p	I_p
Lock detector	ckd ckd₀ ckd₁ ... ckd₇ ... ckd₁₄ ckd₁₅ ... rst Sampling	ckd ckd₀ ckd₁ ... ckd₇ ... ckd₁₄ ckd₁₅ ... rst Sampling
	Sampled ckd₇ = 1 **lock_det=0**	Sampled ckd₇ = 0 **lock_det=1**

Fig. 4.4 *lock_det* and *lock* signals control the CPFD status and charge current to the loop filter

signal goes high. The CPFD operates in the low-jitter mode ($\theta_e < |\theta_M|$) and the turn-on sequence of the charge current transitions from en_{15} back to en_0, resulting in the sequential turn-on current as: $\times 1 \rightarrow \times 2 \rightarrow \times 4 \ldots$, where a low and slowly increasing charge current flows into the loop filter for lower jitter. If it remains in the low-jitter mode for more than 16 clock cycles, then **stab** = 1 and dithering is enabled. Here, the weighting choice of en_0 to en_{15} is designed as 16, 16, 10, 10, 8, 8, 8, 8, 8, 8, 4, 4, 2, 2, 1, and 1.

Based on the control procedure, the simulation waveform of V_{FILT}, the control signals from the fast-lock and low-jitter modes of the PLL is presented in Fig. 4.5. The power-up to **lock** = 1 takes approximately 15 μs and follows the control procedures to reach a stable state then enabling dithering.

4.2.2   Two-Step Switching (TSS) Technique

The TSS technique can reduce current consumption and transient bounce caused by the bonding of the wire and lead frame for the off-chip loop filter. TSS operates by turning on the **UP** or **DN** switches first; CPFD then provides a suitable charge-pump current through the sequence of turning on switches. Because of this sequence, the charge-pump current can be increased smoothly without adding transient noise. The current through CPFD switches is turned off during no-charge or no-discharge periods so as to save power and maintain operation of the current mirror.

In a conventional design, the current mirror provides a constant, steady current for the CP, as demonstrated in Fig. 4.6. Each current-mirror set consumes multiple bias currents to construct a complete CP. With the trade-off between system stability and jitter performance, strong charge-pump currents are usually selected for

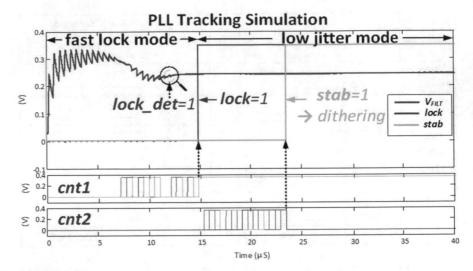

Fig. 4.5 Simulation waveform of the PLL from the powered-up fast-lock mode to the low-jitter mode

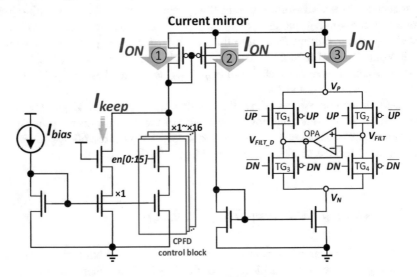

Fig. 4.6 Two-step switching schematic, including CPFD switches

fast-tracking. When the PLL is in the locked state, the current is only required during the short dead-zone period to maintain PLL tracking. When the loop filter is not charged or discharged, the current becomes unnecessarily strong and is wasted on the long t_{REF} period in conventional designs. The proposed TSS, displayed in Fig. 4.7, can save the current in these current-mirror sets only when they are in use. When the loop filter is not charged or discharged, only a low current I_{keep} is maintained in the current mirror. The purpose of I_{keep} is to maintain operation of the

Fig. 4.7 Proposed two-step
switching control can save
99.6% current mirror power

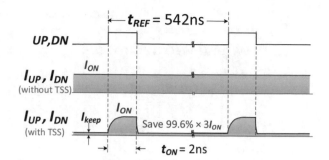

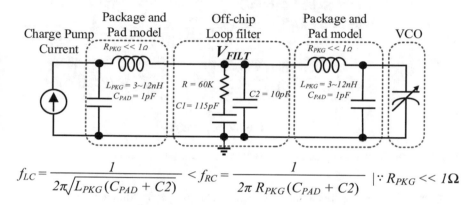

$$f_{LC} = \cfrac{1}{2\pi\sqrt{L_{PKG}(C_{PAD} + C2)}} < f_{RC} = \cfrac{1}{2\pi\,R_{PKG}(C_{PAD} + C2)} \quad |\because R_{PKG} << 1\Omega$$

Fig. 4.8 Equivalent circuit of lumped pad model and off-chip loop filter

current mirror and its quick response to the target current value when *UP* or *DN* is
on. If the turn-on time is $t_{on} = 2$ ns, it will save 99.6% $(1 - t_{on}/t_{REF})$ of the static
power of the current mirror for $t_{REF} = 542$ ns (i.e., 1.843 MHz).

The off-chip loop filter is connected through a lead frame and bonding wire
attached to pads. This lumped pad model consists of bonding wire and a lead frame
as an inductor and the bonding pad as a parasitic capacitor for grounding. The total
inductance is approximately 3–12 nH, depending on the length and material of the
bonding wire and lead frame for different package types, such as QFP, PLCC, or
TSOP. Unfortunately, this connection forms a lumped LC resonator shown in
Fig. 4.8, affecting the transient response when the loop filter is charged or
discharged.

The stability analysis of the PLL with and without the bond wire (pad model) is
shown in Fig. 4.9. The small ac signal goes through the pad model, resulting in only
phase shift without amplitude degradation. The ac response of each pad model
generates 180° phase shift, so there is total 360° phase shift (from input pad to
output pad) for the full system with pad model. The stability analysis of the full-
system magnitude for different corners is almost the same for both with and without
pad model.

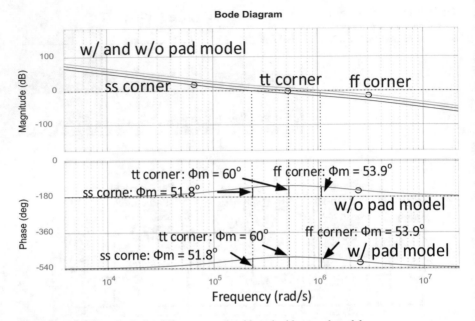

Fig. 4.9 Stability analysis for different corners with and without pad model

Comparison of the transient responses of charging loop filter with and without TSS is shown in Fig. 4.10, where a step function of current with the same magnitude is provided at the time of 0.1 µs. We can find different ringing amplitudes and time periods happening without TSS case. It is because the charge current with high-frequency components of impulse triggers the lumped LC resonator causing oscillation phenomena. However, with TSS, the smooth increasing charge current has no high-frequency components to trigger the same lumped LC resonator. Even without off-chip loop filter, TSS still can reduce power bounce when charge-pump is switching.

Besides the inductor for off-chip loop filter, there are also power and ground inductors. The Fig. 4.11 shows only the power inductor for simplicity. The supply bounce or power/ground noise is proportional to the slew rate (sharpness) of charge-pump current pulse because of L·di/dt effect. The power/ground noise can be alleviated by putting a large part of $C_{decouple}$ near charge-pump. This $C_{decouple}$ is very important to stabilize supply voltage because the VCO is nearby, where the power/ground bounce will degrade VCO's jitter performance because of VCO's intrinsic high-pass characteristic.

This work not only has already put a large $C_{decouple}$ nearby the charge-pump to reduce the supply bond-wire L·di/dt effect, but TSS and smoothly increasing charge current (proposed CPFD) also reduce loop filter bond-wire L·di/dt effect. This synergy effect can reduce both power/ground noise and jitter.

To avoid any timing mismatch that may produce ripple and phase offset leading to high reference spur, the two control paths from PFD to switches of the

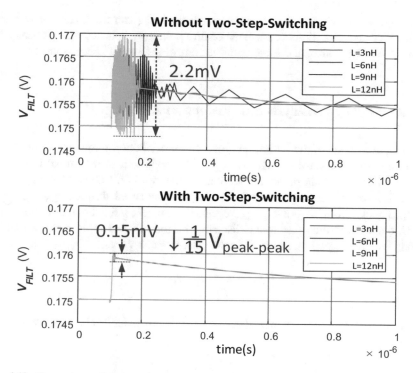

Fig. 4.10 Comparison of the transient responses without and with TSS for off-chip loop filter, including lead frame and bonding wire inductance

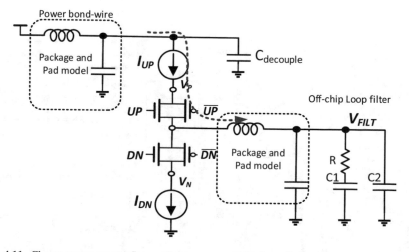

Fig. 4.11 Charge-pump current flow path from power to the loop filter

charge-pump are designed to be matched. Both paths can track each other under different corner conditions because the control path to switches' gates duplicates the same device type, size, and stage with the control path to switches' sources.

4.2.3   Leakage Analysis and Reduction Techniques

The output voltage of the loop filter controls the oscillation frequency of the VCO. Variations in the control voltage correspond to deviations in the output frequency. Figure 4.12 shows a simplified model of all connections on VCO control input node V_{FILT}, including a leakage current. The C_2 is discharged through $I_{LEAKAGE}$ and charged back through I_{UP} every t_{REF} cycle. To maintain a fixed output frequency at VCO, the voltage drop during the discharge period is equal to the voltage buildup during the charge period, resulting in the peak-to-peak voltage ΔV_{FILT} [3].

$$\Delta V_{FILT} = \frac{I_{LEAKAGE} \times t_{discharge}}{C_2} = \frac{I_{UP} \times t_{charge}}{C_2} \qquad (4.2)$$

The power ratio of fundamental reference spur-to-carrier can be expressed as [4]:

$$\left[\frac{P_{spur\_fund}}{P_{carrier}} \right]_{dBc} = 20 \log_{10} \left(\frac{K_{VCO} \times I_{LEAKAGE}}{2\pi C_2 \times f_{REF}^2} \right) \qquad (4.3)$$

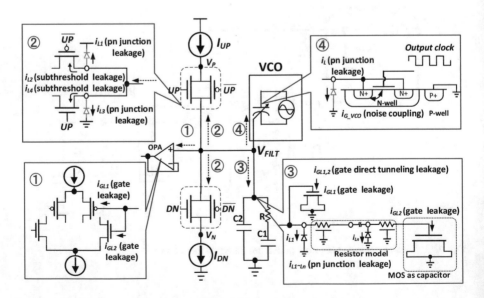

Fig. 4.12 Four possible leakage paths form V_{FILT} node

Fig. 4.13 VCO control input node V_{FILT} and all its connections

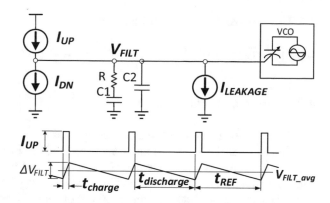

As shown in (4.3), the magnitude of leakage impacts the power of reference spur directly. Careful study of all possible leakage paths from the V_{FILT} node is therefore essential for minimizing or reducing them to attain superior spur performance. The four major leakage paths at the V_{FILT} node are displayed in Fig. 4.13.

Those leakage paths can be classified into four types of leakage mechanisms:

4.2.3.1 Gate Direct Tunneling Leakage

Both paths ① and ③ possess this type of leakage mechanism.

To reduce this leakage, an op-amp design enables the maintenance of a constant W:L ratio, but scales down both W and L values to attain the same transconductance for ①.

The total capacitance of loop filter is 125 pF. Using MOS as a capacitor in the loop filter generally results in a greater area efficiency (half of this chip area), but at the cost of more severe gate leakage (3 μA). For path ③, one option is to use a MOM/MIM capacitor in the chip because, ideally, neither demonstrates a leakage problem, but both enlarge chip area (three times of this chip area). Another option we adopt is to use an external off-chip SMD capacitor to avoid gate leakage (leakage current <0.09 nA).

4.2.3.2 Junction Leakage

Paths ②, ③, and ④ exhibit this type of leakage mechanism.

To minimize this leakage, a smaller device can be used in those circuits and an off-chip resistor can be used for the loop filter instead of the well resistor with heavy junction leakage or poly resistor with larger area because of smaller sheet resistance.

4.2.3.3 Gate-Induced Drain Leakage

Only path ② possesses this type of leakage mechanism.

4.2.3.4 Subthreshold Leakage

Paths ② possess this type of leakage mechanism.

All classified leakage currents are presented in Fig. 4.14 for NMOS with W/L = 10 μm/0.1 μm and a temperature range of 0–120 °C. We can demonstrate that the dominate leakage source is subthreshold leakage, which is almost 10^4 times larger than others. Additionally, the subthreshold current increases with temperature.

After a detailed study of all leakage paths and mechanisms, we focused on reducing the dominate leakage current (i.e., subthreshold leakage current).

The subthreshold leakage current, or I_{sub}, was calculated using the following formula [5]:

$$I_{sub} = I_0 e^{\frac{V_{GS}-V_T}{nU_T}}\left(1 - e^{\frac{-V_{DS}}{U_T}}\right) \tag{4.4}$$

where $I_0 = \frac{W\mu_0 C_{ox} V_T^2 e^{1.8}}{L}$, $U_T = \frac{KT}{q}$ is the thermal voltage, V_T is the threshold voltage, and V_{DS} and V_{GS} are the drain-to-source and gate-to-source voltages, respectively.

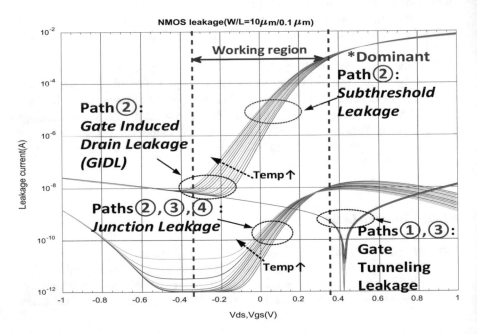

Fig. 4.14 NMOS subthreshold leakage, GIDL, junction leakage, and gate tunneling leakage current

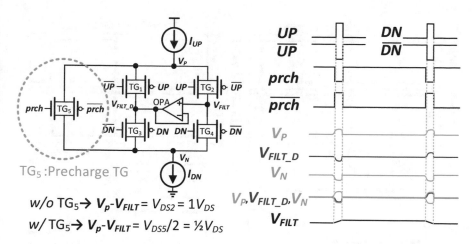

Fig. 4.15 Proposed subthreshold leakage reduction circuit for V_{FILT} node and control waveforms

W and L are the effective transistor width and length, respectively. C_{ox} is the gate oxide capacitance, μ_0 is the carrier mobility, and n is the subthreshold swing coefficient. From this formula, even when V_{GS} is set to 0 V, a current still flows into the channel of the "OFF" MOS transistor because of drain and source terminals across with V_{DS}. In all variables listed in Eq. (4.4), I_0 and U_T are physical and process parameters that are difficult to change during circuit operation. We can discern that greater V_{DS} corresponds with greater I_{sub} by evaluating this formula and focusing solely on the last term. I_{sub} can be minimized or even approximately equal to zero by setting $V_{DS} \approx 0$. One method for reducing V_{DS} is enlarging each TG (transmission gate), but not only will the switching noise couple with the current source through the large C_{gs}, but efficiency will be reduced because two TGs are in series between V_P and V_N. A preferable method for reducing V_{DS}, as proposed in Fig. 4.15, is adding a precharge switch, TG$_5$, between V_P and V_N to directly reduce its voltage drop, resulting in half of the original V_{DS} values on TG$_1$ and TG$_3$.

The circuit operation is as follows: the precharge signal is the inverse of **UP** OR **DN** (i.e., $prch = \overline{UP + DN}$). When **UP** or **DN** goes high, **prch** will go low for normal tracking procedures. Otherwise, V_{FILT} remains constant because both **UP** and **DN** remain low on the right-hand path, and **prch** is high resulting in the shortening of V_P to V_N through TG$_5$ on the left-hand path. In addition to this, two other things also happen on the left-hand path: V_P shortened to V_{FILT_D} and V_N shortens to V_{FILT_D} through TG$_1$ and TG$_3$; at the same time, V_{FILT_D} has the same voltage as V_{FILT} from the unit-gain buffer. The result is $V_P \approx V_{FILT} \approx V_N \approx V_{FILT}$, meaning that when $V_{DS} \approx 0$, the possible leakage path is cut from the oscillator input of V_{FILT}. Finally, the V_{FILT} node exhibited almost no subthreshold leakage. The control waveforms are displayed in Fig. 4.15.

In addition to those DC leakage currents, this was also the case with the AC coupling noise, which was the gate tunneling current from the multiphase VCO output through the N-Well of varactor to the V_{FILT} node. This noise, however, can

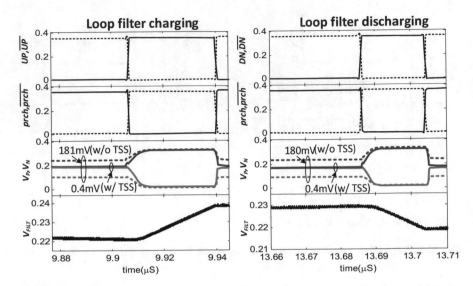

Fig. 4.16 Simulation of subthreshold leakage reduction

be neglected by the long-term average effect of clock output signals with a symmetrical amplitude waveform.

Because of the TSS technique described in the previous section, the source of currents I_{UP} and I_{DN} is kept at a low value of I_{keep} to save power, and the I·R voltage drop in TG$_5$ is less extreme than that of conventional CPs and approaches zero. The simulation waveform demonstrates that the voltage drop in TG$_5$ is from 181 mV (without TSS and TG$_5$) down to approximately 0.4 mV (with TSS) in Fig. 4.16, which reduces the $(1 - \exp.(-V_{DS}/V_T))$ term to 1/65 of original subthreshold current from the sensitive node V_{FILT} and improves spur level for about -36 dB by (3).

4.2.4   Dither Technique

The reference spur for a phase-locked loop is generated by the periodic ripples on the control line of the VCO [6]. The magnitude of the reference spur can be calculated from Eq. (4.3):

$$\left[\frac{P_{spur\_fund}}{P_{carrier}}\right]_{dBc} = 20\,log_{10}\left(\frac{K_{VCO} \times \Delta V_{FILT}}{2\pi f_{REF}}\right) \tag{4.5}$$

From Eq. (4.5), the spur amplitude can be improved with lower K_{VCO} and ΔV_{FILT} and larger f_{REF}. The first item of the smaller K_{VCO} can be achieved with fine tuning through control of the varactor. The second item, the smaller ΔV_{FILT}, can be minimized by using the previous leakage analysis and proposed precharge circuit.

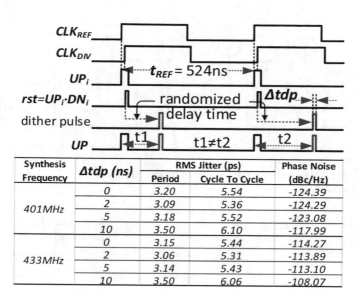

| Synthesis | Δtdp (ns) | RMS Jitter (ps) | | Phase Noise |
Frequency		Period	Cycle To Cycle	(dBc/Hz)
	0	3.20	5.54	-124.39
401MHz	2	3.09	5.36	-124.29
	5	3.18	5.52	-123.08
	10	3.50	6.10	-117.99
	0	3.15	5.44	-114.27
433MHz	2	3.06	5.31	-113.89
	5	3.14	5.43	-113.10
	10	3.50	6.06	-108.07

Fig. 4.17 Dither technique randomly adds an extra pulse into the fixed reference period and the relationship of dither pulse width with jitter and phase noise

The last item, the increased f_{REF} is a trade-off between the commercially available channel number and crystal oscillator frequencies. We proposed that the dither technique does not directly increase f_{REF} but instead causes an equivalent increase in f_{REF} and randomizes the main reference spur tone at f_{REF}.

The concept of the proposed dither technique is to randomly add an additional pulse into the fixed reference period, causing the spur tone to average out in the frequency domain. The randomly delayed pulses are initially triggered by the **rst** pulse and then generated through a series of stages with different delay times. Figure 4.17 presents the concept of randomizing the regular period of the reference spur, where $t1 \neq t2 \neq T_{REF}$. Since the dither technique increases the on time of the charge current, the noise contribution from charge-pump is increased. The 2-ns dither pulse width is chosen to optimize jitter and phase noise performance.

Figure 4.18 presents the block diagram of the dither technique, where the circuit is triggered by an **rst** pulse and the ***dither_en*** signal enables the linear feedback shift register (LFSR) to generate a pseudo-random sequence code (***d[0:7]***) to control the current-controlled delay cells (CCDC). From Fourier analysis, 8 bit random code can reduce −48 dB reference spur [7]. However, since the total maximum delay time can only reach half the reference period of 270 ns, the expected reference spur reduction is only −24 dB.

The dither technique not only averages out the reference spur tone, but also reduces the peak-to-peak ripple amplitude of V_{FILT}. Simulation waveforms are shown in Fig. 4.19, where the ΔV_{FILT} value is reduced from 0.24 mV to 0.15 mV, which is an approximately 38% reduction, resulting in further −4 dBc reference spur reduction based on (5).

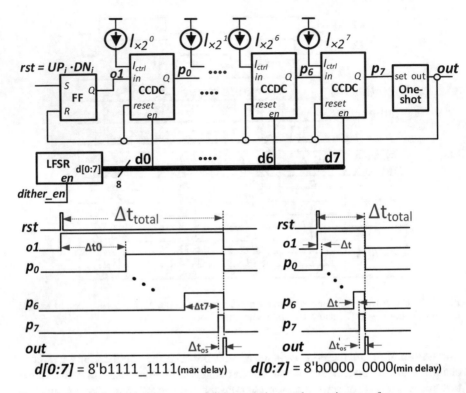

Fig. 4.18 Block diagram of the proposed dither technique and operation waveforms

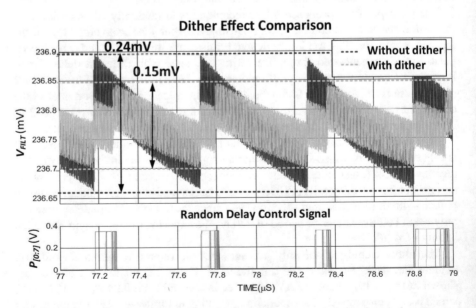

Fig. 4.19 Dither effect comparison and randomized delay time control signal on each stage

4.3 Circuit Implementation

Based on the proposed design techniques, a 0.35-V, low-power frequency synthesizer was designed for implantable applications. Most of the concepts and circuits of the proposed techniques were detailed in the previous section.

4.3.1 Low Phase Noise Wide-Tuning-Range VCO

A VCO is one of the key components in a PLL, and it significantly affects the overall performance of PLLs. In this chapter, by employing a multiphase scheme to implement the VCO, we were able to simultaneously achieve a higher operating frequency and wider tuning range. The low-voltage differential-type multiloop ring oscillator with dual current-varactor tuning architecture is presented in Fig. 4.20(a). The coarse tune is designed to control the bias current of delay cells to calibrate the corner frequency to be close to the synthesis frequency during the power-up sequence [8]. The fine tune is achieved through the linear C-V tuning curve of varactors and its range is depicted in Fig. 4.21 for different corners and corresponding control codes. Its marked linearity can be observed in all tuning ranges with a K_{VCO} of approximately 128 MHz/V.

The phase noise of the VCO is affected by two major noise sources, one is uncorrelated noise source, which is device noise generated by thermal noise in the devices, and the other is correlated noise source, which is power/substrate noise and generally larger than those from device noise sources in high-speed mixed analog–digital integrated circuits [9].

The schematic of the VCO delay cell is shown in Fig. 4.20(b), which is the same as that in [10], except replacing **Vctrl** by ground and adding varactors as output loading for fine tuning control. The benefit of the varactor at the output node is not only its linearity, but also reduces the sensitivity of noise caused by voltage-dependent capacitance of output drain junctions [11].

To minimize the uncorrelated noise, the tuning of delay cell circuit follows Hajimiri's suggestions [12, 13]: (1) A higher signal swing ΔV implies a higher signal power P and results in a greater signal-to-noise ratio [12]. (2) The quick and symmetric rising and falling of the edge results in less upconversion noise [13].

To minimize the correlated noise, layout is maintained in a symmetrical fashion [14] and on-chip bypass capacitors 34 pF is placed in close proximity to output buffers [15]. The analog and digital circuits are separated by at least four times the thickness of the epitaxial layer, so the resistance between the substrate contacts is independent of their separation [16]. The epitaxial layer of this technology node is approximately 1 μm, whereas the distance between the VCO and the other circuits is greater than 10 μm, as shown in Fig. 4.24.

A latch-type waveform-sharpening mechanism was added to avoid signal degradation in the transfer path. Compared with a simple buffer stage, not only does the

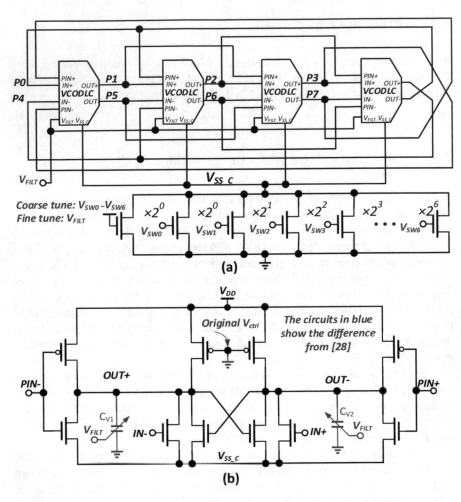

Fig. 4.20 (**a**) Proposed dual current-varactor tuning architecture VCO and (**b**) VCO delay cell

VCO output slew rate shape, but sensitivity to common-mode noises is also reduced. The simulation waveform is presented in Fig. 4.22, which displays a large swing in signal with the symmetric rising and falling edges of the VCO output, as suggested.

Figure 4.23 presents the simulation of VCO phase noise by adding 20-mV sine-wave noise to the power and ground lines to verify the contribution of supply noise to VCO phase noise. The simulation results reveal a 28–33 dBc/Hz degrade at 1 MHz offset, indicating that VCO is more sensitive to ground noise than it is to power noise. The simulation also demonstrated that replacing wave sharpening with a normal buffer will degrade phase noise at a low offset frequency. This indicates that enhancing the clock slew rate can improve the immunity to low-frequency flicker noise.

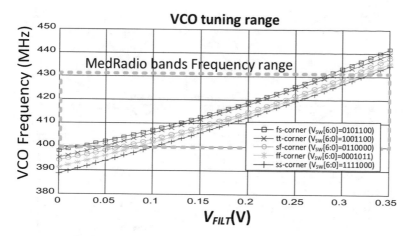

Fig. 4.21 VCO fine tune range for different corners

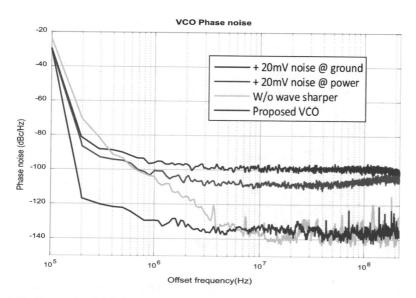

Fig. 4.22 Phase noise simulation for supply noise, w/ and w/o wave sharper

4.4   Measurement Results

The proposed PLL was fabricated using 90-nm CMOS technology with low V_T devices and mounted on a FR4 board for measurement. Figure 4.24 presents die microphotograph and power consumption. The chip area is approximately 0.275 mm × 0.14 mm, excluding the pad area. The reference frequency was generated from a crystal oscillator, and its amplitude was reduced to 0.35 V by using a resistive ladder.

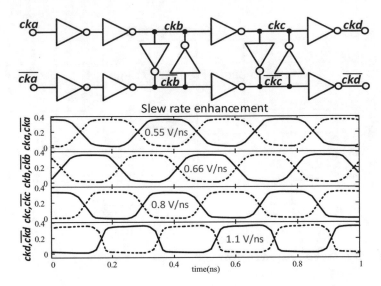

Fig. 4.23 Waveform sharpening improves the stage-by-stage slew rate in the transfer path

Power Breakdown of the PLL	
Total Power @ 0.35V	238μW
PFD	0.2μW
CP	7.9μW
VCO	132μW
Divider	49.4μW
Controller	12.3μW
Buffer*	36μW

*:Including input and output
clock buffers except OCD

Fig. 4.24 Die microphotograph and power consumption of the proposed frequency synthesizer

Figure 4.25 displays the measured step response; the lock time was approximately 20 μs, which is remarkably shorter than 90 μs [17] for the same reference frequency without CPFD.

Figure 4.26 displays the measured jitter histogram when the output frequency was set to 433.15 MHz. The output peak-to-peak and RMS jitter were 42.7 ps and 5.3 ps, respectively.

Because of leakage reduction and dither techniques, the reference-spur tone was averaged out in the frequency domain. Figure 4.27 depicts the measured frequency spectrum of the PLL, with the measured reference-spur tone at 1.832-MHz frequency offset demonstrating a maximum power of −55.48 dBc.

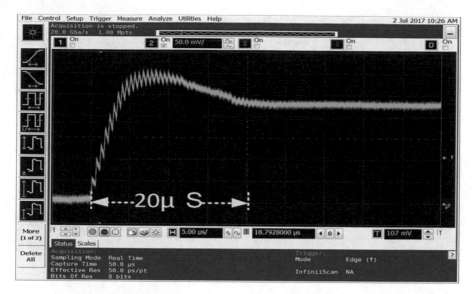

Fig. 4.25 Measured step response

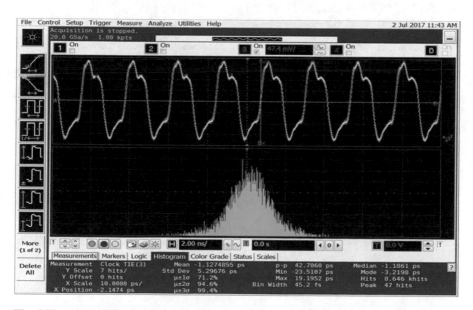

Fig. 4.26 Measured jitter performance

Figure 4.28 presents the measured phase noise of the PLL. At 1-MHz frequency offset, the measured phase noise was −105.7 dBc/Hz. Table 4.2 summarizes the performance of the proposed PLL and comparison with prior works. The FOMs of this PLL are favorably compared to the other previously published works.

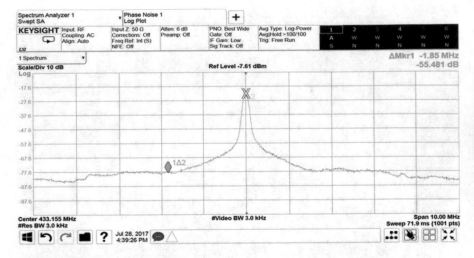

Fig. 4.27 Measured frequency spectrum

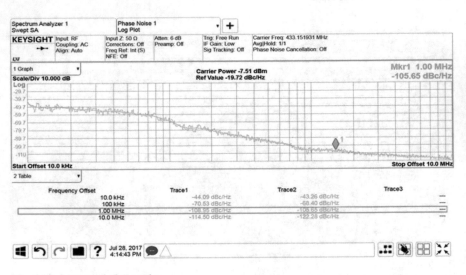

Fig. 4.28 Measured phase noise

Table 4.2 PLL performance comparison

	TCAS-I 2011 [18]	JSSC 2012 [17]	TCAS-II 2014 [19]	ISSCC 2014 [20]	ISSCC 2016 [21]	ISSCC 2017 [22]	This work
Technology (CMOS)	90 nm	130 nm	65 nm	20 nm	40 nm	28 nm	90 nm
Supply voltage (V)	0.5	0.5	0.4	0.9	1.1	0.8	0.35
Area (mm^2)	0.074	0.0736	0.0081	0.012	0.0216	0.049	0.0385
Power (mW)	0.4	0.44	0.109	3.1	2.915	2.5	0.24
Locking time (μs)	N/A	90	N/A	N/A	N/A	N/A	20
Output freq. (GHz)	0.4	0.4–0.433	0.35	1.6	2.4	1	0.4–0.433
Reference spur (dBc)	N/A	−38.3	−55.3	N/A	−30.65	N/A	−55.48
Jitter (ps)	77.78 (pk-pk) 9.62 (rms)	49.1 (pk-pk) 5.5 (rms)	30.8 (rms)	49.87 (pk-pk) 5.89 (rms)	30.85 (pk-pk) 3.54 (rms)	3 (rms)	42.7 (pk-pk) 5.3 (rms)
Jitter/cycle (%)	3.8	2.4	N/A	N/A	N/A	N/A	2.3
Phase noise @1 MHz (dBc/Hz)	−87	−91.5	−90	N/A	−101	N/A	−105.65
FOM[a]	25.6	4.5	2.61	658.04	49.92	420.39	1
FOM$_{Jitter}$ (dB)[b]	−224.3	−228.8	−219.9	−219.7	−224.4	−226.5	−231.7

[a][23]:
$$FOM = \left[\frac{area(mm^2)}{\left(\frac{tech}{0.09}\right)^2}\right] \left[\frac{mW}{Mhz}\right]^{1.5} \left[Jitter_{rms(ps)} \sqrt{mW}\right]^2$$
[39] : FOM =
$$\left[\frac{area(mm^2)}{\left(\frac{tech}{0.09}\right)^2}\right] \left[\frac{mW}{Mhz}\right]^{1.5} \left[Jitter_{rms(ps)} \sqrt{mW}\right]^2$$

[b][24]: $FOM_{Jitter} = 10 \log \left[\sigma_{jitter}^2 \frac{P_{DC}}{1\,mW}\right]$

References

1. Yang, C., & Liu, S. (2000). Fast-switching frequency synthesizer with a discriminator-aided phase detector. *IEEE Journal of Solid-State Circuits, 35*(10), 1445–1452.
2. Gardner, F. (1980). Charge-pump phase-lock loops. *IEEE Transactions on Communications, 28*(11), 1849–1858.
3. Azarian, M., & Ezell, W. (2013). *A simple method to accurately predict PLL reference spur levels due to leakage current.* Linear Technology.
4. Gao, Z. Q., Lan, J. B., Liu, X. W., & Yin, L. (2014). The delta-sigma modulator of fractional-N frequency synthesizer for wireless sensors network applications. *Key Engineering Materials, 609*, 1014–1019.
5. Butzen, P. F., & Ribas, R. P. (2006). *Leakage current in sub-micrometer cmos gates* (pp. 1–28). Universidade Federal do Rio Grande do Sul.
6. Razavi, B., & Behzad, R. (1998). *RF microelectronics* (Vol. 1). Prentice Hall.

7. Liang, C.-F., Chen, H.-H., & Liu, S.-I. (2007). Spur-suppression techniques for frequency synthesizers. *IEEE Transactions on Circuits and Systems II: Express Briefs, 54*(8), 653–657.
8. Wilson, W. B., Moon, U.-K., Lakshmikumar, K. R., & Dai, L. (2000). A CMOS self-calibrating frequency synthesizer. *IEEE Journal of Solid-State Circuits, 35*, 1437–1444.
9. Heydari, P. (2004). Analysis of the PLL jitter due to power/ground and substrate noise. *IEEE Transactions on Circuits and Systems I: Regular Papers, 51*, 2404–2416.
10. Liu, H. Q., Siek, L., Goh, W. L., & Lim, W. M. (2008). A 7-GHz multiloop ring oscillator in 0.18-μm CMOS technology. *Analog Integrated Circuits and Signal Processing, 56*(3), 179–184.
11. Herzel, F., & Razavi, B. (1999). A study of oscillator jitter due to supply and substrate noise. *IEEE Transactions on Circuits and Systems II: Analog and Digital Signal Processing, 46*(1), 56–62.
12. Hajimiri, A., & Lee, T. H. (1999). *The design of low noise oscillators*. Springer Science & Business Media.
13. Hajimiri, A., Limotyrakis, S., & Lee, T. H. (1999). Jitter and phase noise in ring oscillators. *IEEE Journal of Solid-State Circuits, 34*, 790–804.
14. Blalack, T., Lau, J., Clément, F. J., &Wooley, B. A. (1996). Experimental results and modeling of noise coupling in a lightly doped substrate. In *Electron Devices Meeting, 1996. IEDM'96., International*, pp. 623–626.
15. Heydari, P., & Pedram, M. (2003). Ground bounce in digital VLSI circuits. *IEEE Transactions on Very Large Scale Integration (VLSI) Systems, 11*, 180–193.
16. Su, D. K., Loinaz, M. J., Masui, S., & Wooley, B. A. (1993). Experimental results and modeling techniques for substrate noise in mixed-signal integrated circuits. *IEICE Transactions on Electronics, 76*, 760–770.
17. Chen, W.-H., Loke, W.-F., & Jung, B. (2012). A 0.5-V, 440-μW frequency synthesizer for implantable medical devices. *IEEE Journal of Solid-State Circuits, 47*, 1896–1907.
18. Cheng, K.-H., Tsai, Y.-C., Lo, Y.-L., & Huang, J.-S. (2011). A 0.5-V 0.4–2.24-GHz inductorless phase-locked loop in a system-on-chip. *IEEE Transactions on Circuits and Systems I: Regular Papers, 58*, 849–859.
19. Moon, J.-W., Choi, K.-C., & Choi, W.-Y. (2014). A 0.4-V, 90 350-MHz PLL with an active loop-filter charge pump. *IEEE Transactions on Circuits and Systems II: Express Briefs, 61*, 319–323.
20. Liu, J., Jang, T.-K., Lee, Y., Shin, L., Lee, S., Kim, T., et al. (2014). 15.2 A 0.012 mm^2 3.1 mW bang-bang digital fractional-N PLL with a power-supply-noise cancellation technique and a walking-one phase selection fractional frequency divider. In *2014 IEEE International Solid-State Circuits Conference (ISSCC), 2014*, pp. 268–269.
21. Yeh, C.-W., Hsieh, C.-E., & Liu, S.-I. (2016). 19.5 A 3.2 GHz digital phase-locked loop with background supply-noise cancellation. In *Solid-State Circuits Conference (ISSCC), 2016 IEEE International, 2016*, pp. 332–333.
22. T. Jang, S. Jeong, D. Jeon, K. D. Choo, D. Sylvester, and D. Blaauw, "8.4 A 2.5 ps 0.8-to-3.2 GHz bang-bang phase-and frequency-detector-based all-digital PLL with noise self-adjustment," in Solid-State Circuits Conference (ISSCC), 2017 IEEE International, 2017, pp. 148–149.
23. Fahim, A. M. (2003). A compact, low-power low-jitter digital PLL. In *29th European Solid-State Circuits Conference (ESSCIRC'03), 2003*, pp. 101–104.
24. Gao, X., Klumperink, E. A., Geraedts, P. F., & Nauta, B. (2009). Jitter analysis and a benchmarking figure-of-merit for phase-locked loops. *IEEE Transactions on Circuits and Systems II: Express Briefs, 56*(2), 117–121.

Chapter 5
Introduction of ADC

This chapter will introduce the basic working principle and simulation of the analog-to-digital converter (ADC). ADCs have different requirements for speed and accuracy in different applications, and power consumption needs to be considered. Next, the basic operation steps and principles of the ADC will be introduced, as well as some of its important static and dynamic performance indicators. Furthermore, the limitations brought about by the environment, including timing jitter and thermal noise limitations, will be discussed. Common ADC architectures, including FLASH, pipeline, successive approximation, and sigma-delta data converters, will also be presented. In this chapter, we will focus on the successive approximation of ADC algorithms, calibration techniques, and the latest development trends and discuss ADC simulation and testing, as well as some basic knowledge of FFT using digital signal processing and frequency selection and leakage issues. Finally, we will investigate the commonly used test methods of Nyquist ADC.

5.1 ADC Performance and Application

Analog-to-digital converters are an indispensable part of modern electronic systems. For every mixed-signal chip or any system-on-chip we fabricate, data converters are the bridge between them and the outside world. At present, since signal processing is actually carried out in the digital domain, the signal must be digitized through an ADC. For example, in biomedical applications, they are the key link between the front-end analog sensor and the back-end digital computer, realizing the functions of monitoring, judging, and recording biomedical signals. Depending on the application, the ADC is the interface between the real world (continuous time and amplitude) and the virtual world (discrete time and amplitude). Therefore, whether you are

C.-C. Hung, S.-H. Wang, *Ultra-Low-Voltage Frequency Synthesizer and Successive-Approximation Analog-to-Digital Converter for Biomedical Applications*, Analog Circuits and Signal Processing, https://doi.org/10.1007/978-3-030-88845-9_5

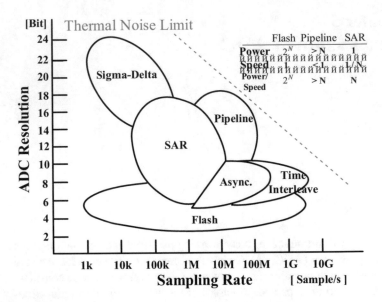

Fig. 5.1 Comparison of ADC resolution bits and sampling rate performance

an ADC designer or a system designer, you should understand the architecture and limitations of the ADC.

ADCs can be used in different fields. In fact, different applications have different requirements for ADCs. For example, radar requires an ADC with a very high frequency and a low resolution, or a digital voltmeter requires an ADC with a very high resolution and a low conversion rate. Figure 5.1 shows the comparison of ADC resolution bits and sampling rate performance [1]. The sigma-delta frequency is lower, but the resolution is higher, while the flash frequency is higher, but the resolution is lower. Pipeline and SAR are somewhere in between.

5.1.1 Nyquist Rate ADC Performance Comparison

We further compared the performance of the Nyquist rate ADCs, including flash, pipeline, and SAR between power and speed. The power of flash ADC is usually 2^N, where N denotes the number of bits, because there is a large number of comparators working at the same time. The power of the pipeline is greater than N because there are many operational amplifiers working at the same time. The SAR has only a comparator, so its power is the smallest compared to others. But the speed of flash ADC is very high, whereas that of SAR is low. In a sense, we can imagine the pipeline having N SARs performing ADC work at the same time, and we have already paid the price of power consumption for speed. Or we can imagine SAR processing only one ADC at a time, and it takes N times to complete the work of the pipeline. We pay the price of speed for power consumption.

5.1.2 Data Conversion

A/D conversion means that we need to convert analog signals to digital representations. The typical A/D conversion process is as follows: The input signal is sampled through an appropriate anti-aliasing filter. Then, a discrete-time signal is generated and then quantized to obtain a signal with discrete amplitude levels, as presented in Fig. 5.2. Thus, when people say "A/D converters," they basically mean the entire operation, including sampling, quantization, and coding.

When it comes to sampling, Nyquist's requirements need to be considered: the maximum input signal frequency ϕ 1/2 sampling frequency. As can be seen from Fig. 5.3, the green signal is 1 Hz. Using a 4-Hz sampling rate is sufficient to reconstruct the signal. However, the 3-Hz signal cannot be distinguished due to the use of 4-Hz sampling, and the red signal actually violates Nyquist's requirements. Thus, it cannot be distinguished or reconstructed. It is either we must ensure that the signal is band-limited or we must use an anti-aliasing filter to limit the signal to a specific frequency range.

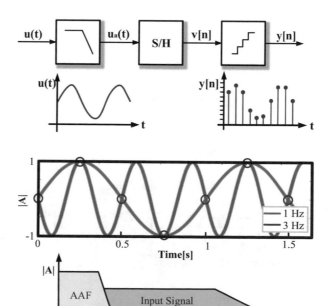

Fig. 5.2 Analog-to-digital conversion

Fig. 5.3 Sampling may cause aliasing

5.1.2.1   ADC Operation: Sampling

The first process in ADC is to sample the analog signal and maintain the input voltage on the capacitor. This can be performed in two different ways, namely, bottom-plate sampling and top-plate sampling, as shown in Fig. 5.4(a). The difference between the two technologies is the need to connect the sample-and-hold circuit to the bottom or top plate of the capacitor array. Although the top-plate sampling technique may have problems of charge injection and parasitic capacitance, for low power consumption and low power-supply voltage, it is recommended for the ADC. This is because the number of switches required for top-plate sampling is less than that of the bottom-plate sampling. In either case, the root mean square (RMS) noise is kT/C due to the unavoidable thermal noise associated with the sampling switch.

5.1.2.2   ADC Operation: Quantization

Once the signal is sampled, it is quantized. The quantizer has the input and output characteristics shown in the black step line in Fig. 5.4(b).

Although it is a nonlinear characteristic, ideally, the A/D converter encodes the continuous-time analog input voltage V_{in} into a series of discrete N-bit digital words, satisfying the following relationship:

$$V_{IN} = V_{FS} \sum_{k=0}^{N-1} \frac{b_k}{2^{k+1}} + \epsilon \qquad (5.1)$$

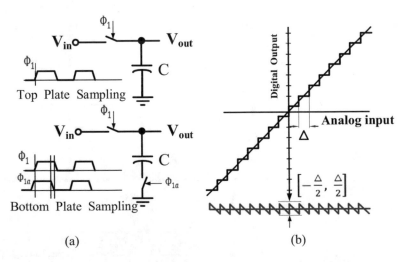

(a)   (b)

Fig. 5.4 ADC operation: (**a**) sampling technique, (**b**) quantization and quantization error

V_{FS} is the full-scale voltage, b_k is each output bit, and ϵ is the quantization error. This relationship can also be expressed in terms of least significant bit (LSB) or quantum voltage level, as shown below:

$$\Delta = \frac{V_{FS}}{2^N} = V_{LSB} \tag{5.2}$$

$$V_{IN} = \Delta \sum_{k=0}^{N-1} b_k 2^k + \epsilon \tag{5.3}$$

Since the quantization error is a complex nonlinear function of the input, there are usually three assumptions: (1) it is a signal similar to noise, (2) the noise contained is independent of the input, and (3) the quantization noise is white noise in the spectrum. Based on these assumptions, the probability density function is consistent from $-\Delta/2$ to $+\Delta/2$, where Δ denotes the step size.

5.1.2.3   The Signal-to-Noise Ratio of an Ideal Quantizer

If we want to know the SNR of an ideal data converter, first, we must know the magnitude of the signal and noise separately. Because the input is a sine wave, the power and RMS voltage of the signal are expressed as follows:

$$P_s = \frac{V_{FS}^2}{8} \tag{5.4}$$

$$V_{s\sigma} = \frac{1}{2} \frac{V_{FS}}{\sqrt{2}} = \frac{2^N \Delta}{2\sqrt{2}} \tag{5.5}$$

where V_{FS} denotes the peak-to-peak value of the digital input signal (full scale). Assuming that the quantization error voltage is uniformly distributed from the code width $-\Delta/2$ to $+\Delta/2$, the RMS quantization noise voltage is expressed as follows:

$$V_{Q\sigma} = \frac{\Delta}{\sqrt{12}} = \frac{V_{FS}}{2^N \sqrt{12}} = \frac{V_{FS}/2}{2^N \sqrt{3}} \tag{5.6}$$

For full-scale, single-tone sine wave input, the SNR is expressed as follows:

$$SNR = \frac{V_{s\sigma}}{V_{Q\sigma}} = 2^N \sqrt{\frac{3}{2}} \tag{5.7}$$

$$\text{SNR}_{\text{dB}} = 20 \log \left[2^N \sqrt{\frac{3}{2}} \right] = N \cdot 20 \log (2) + 10 \log \left(\frac{3}{2} \right)$$

$$= 6.02N + 1.76 \text{ dB} \tag{5.8}$$

If the N-bit ADC is completely uniformly quantized, the above formula expresses the SNR in dB. If the SNR is obtained from the measurement result, it includes not only the SNR but also all possible distortions and harmonics, so what is actually measured is the SNDR (D stands for distortion). The expression to obtain the effective number of bits (ENOB) is as follows:

$$\text{ENOB} = \frac{\text{SNDR}_{\text{dB}} - 1.76}{6.02} \tag{5.9}$$

5.1.3   Performance Index

5.1.3.1   Static Performance Index

The static performance index is the A/D converter specification that can be tested at a low speed or even at a constant voltage. These specifications include offset error, gain error, differential nonlinearity (DNL), and integral nonlinearity (INL), as presented in Fig. 5.5.

Offset error: The deviation when the behavior of the A/D converter is zero. The first switching voltage should be 1/2 LSB higher than the analog ground. The offset error is the deviation of the actual conversion voltage from the ideal 1/2 LSB, as shown by AOS in Fig. 5.5(a). The offset error can be easily adjusted through calibration.

Gain error: It refers to the deviation between the slope of the zero-scale and full-scale line passing through the endpoint of the A/D converter and the ideal slope of $2^N/V_{\text{FS}}$ code per volt presented in Fig. 5.5(b). Like offset errors, gain errors can be easily corrected through calibration.

The nonlinearity presented in Fig. 5.5(c) can be expressed by DNL and INL.

(DNL): It is the deviation of the code conversion width from the ideal width of 1 LSB presented in Fig. 5.5(d). The width of all codes in an ideal A/D converter is 1 LSB, so the DNL for each code should be zero. The DNL can be greater than 1 LSB, but cannot be less than −1 LSB. If the DNL is −1 LSB, the code is lost.

INL: It is the distance between the code center and the ideal line in the characteristics of the A/D converter presented in Fig. 5.5(e). If all code centers fall under the ideal line, then the INL of each code should be zero. According to the definition of the "ideal line," there are two possible ways to express the maximum INL: the end point INL and the best-fit straight-line INL. The number of INL in the data sheet can

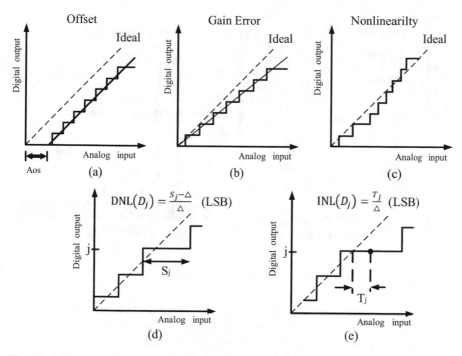

Fig. 5.5 Static performance index: (**a**) offset error, (**b**) gain error, (**c**) nonlinearity, (**d**) differential nonlinearity, (DNL) (**e**) integral nonlinearity (INL)

be reduced by referring to the "best-fit straight-line" line instead of the maximum INL in the ideal line.

5.1.3.2 Dynamic Performance Index

Dynamic performance is usually characterized by the SNR.

SNR: It is the ratio of the signal power (usually a full-scale sine wave) to the total noise generated by the ideal quantizer and circuit noise. The SNR calculates the noise in the entire Nyquist interval.

Signal-to-noise ratio + distortion ratio (SINAD or SNDR): It is similar to the SNR but also takes into account the nonlinear distortion items generated by the input sine wave. SINAD is the ratio of the RMS of the signal to the mean value of the root-sum-square (RSS) of all other spectral components, including harmonics, but excluding DC.

Dynamic range: It is the value of the input signal when the SNR (or SINAD) is 0 dB. This parameter is useful when some data converters cannot obtain the maximum SNR (or SINAD) under 0-dB full-scale input. Figure 5.6 demonstrates that the peak values of the SNR and SNDR are 62 and 76 dB, respectively, and the dynamic range is 80 dB [2].

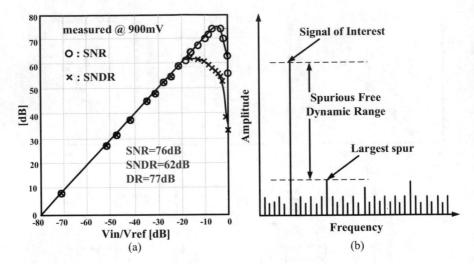

Fig. 5.6 SNR, SNDR, dynamic range, and SFDR

ENOB: It is the number of bits used to measure the SNR and distortion ratio. Intuitively speaking, this makes sense, because every time the resolution increases by 1 bit, the step size can be reduced by 1/2, that is, the noise power is reduced by 1/4, which is an increase of 6 dB. The link between SNDR and ENOB is expressed in dB as follows:

$$\text{ENOB} = \frac{\text{SNDR}_{dB} - 1.76}{6.02} \tag{5.10}$$

Spurious-free dynamic range (SFDR): It is the ratio of the RMS signal amplitude to the RMS value of the highest spurious spectral component in the first Nyquist zone. The SFDR provides information similar to total harmonic distortion but focuses on the worst tone. It depends on the input amplitude. For larger input signals, the highest pitch is given by one of the harmonics of the signal. For input amplitudes far below the full scale, due to the nonlinear characteristics of the converter, the distortion caused by the signal is negligible, and other tones not generated by the input will dominate.

In most practical applications, the ADC input is a frequency band (some inevitable system noise will always be added), so the quantization noise is often random. However, in spectrum analysis applications (or when performing an FFT on an ADC using a spectrally pure sine wave), the correlation between the quantization noise and the signal depends on the ratio of the sampling frequency to the input signal. To accurately measure the harmonic distortion of the ADC, it is important to ensure that the FFT size is large enough, and the frequency ratio must be selected appropriately to ensure that the distortion of the ADC can be measured and the resulting distortion is not related to the quantization noise. Figure 5.7 presents the FFT output of an ideal 12-bit ADC. Note that the average noise floor of the FFT is about 100 dB lower than

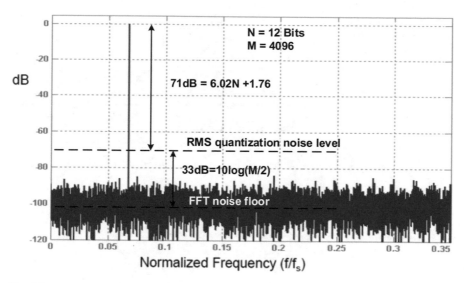

Fig. 5.7   Estimating ENOB from the spectrum overview

the full scale, but in theory, the standard SNR of a 12-bit ADC is 74 dB. The noise floor of the FFT is not the SNR of the ADC, because the function of the FFT is similar to that of an analog spectrum analyzer with a bandwidth of fs/M, where M denotes the number of points in the FFT. Therefore, due to the FFT processing gain, the theoretical FFT noise floor is $10 \cdot \log_{10}(M/2)$ dB lower than the quantized noise floor. If it is an ideal 12-bit ADC with an SNR of 74 dB, a 4096-point FFT will result in a processing gain of $10 \cdot \log_{10}(4096/2) = 33$ dB, resulting in a total FFT noise floor of $74 + 33 = 107$ dBc. Since the actual SNR is 71 dB, ENOB is 11.5 bits.

5.1.4   Limitations of the Analog-to-Digital Converter Architecture

5.1.4.1   Sinusoidal Input Timing Jitter Limit

The most basic dynamic characteristic of sample and hold is that the signal input to the buffer amplifier can be quickly sampled to the hold capacitor, as presented in Fig. 5.8. The short (but nonzero) interval required for this operation is called the aperture time (or sampling aperture). Therefore, when the signal is not moving, the clock input will be affected by noise and cause timing jitter, thus affecting the result of ADC SNR.

The following simple analysis can be used to predict the impact of aperture and sampling clock jitter on the ideal ADC SNR. Suppose the input signal is

Fig. 5.8 Limitation caused
by jitter in the sinusoidal
input timing

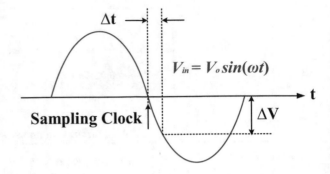

Fig. 5.8 Limitation caused by jitter in the sinusoidal input timing

$$V_{in} = V_o \sin(\omega t) \tag{5.11}$$

The transient rate of this signal is given as follows:

$$dv/dt = \omega V_o \cos(\omega t) \tag{5.12}$$

The RMS value of dv/dt can be obtained by dividing the amplitude ωV_o by $\sqrt{2}$:

$$dv/dt|_{rms} = \omega V_o/\sqrt{2} \tag{5.13}$$

Now, let Δv_{rms} = RMS voltage error and Δt_{rms} = RMS aperture jitter and substitute:

$$\Delta v_{rms}/\Delta t_{rms} = \omega V_o/\sqrt{2} \tag{5.14}$$

Calculate Δv_{rms}:

$$\Delta v_{rms} = \Delta t_{rms}\omega V_o/\sqrt{2} \tag{5.15}$$

The RMS value of the full-scale input sine wave is $V_o/\sqrt{2}$, so the ratio of the RMS signal to the RMS noise is

$$SNR = 20\log\left[\frac{V_o/\sqrt{2}}{\Delta v_{rms}}\right] = 20\log\left[\frac{V_o/\sqrt{2}}{\Delta t_{rms}\omega V_o/\sqrt{2}}\right] = 20\log\left[\frac{1}{\Delta t_{rms}\omega}\right]$$

$$= 20\log\left[\frac{1}{2\pi f \Delta t_{rms}}\right] \tag{5.16}$$

From the relationship between the SNR and input frequency, the requirements for clock jitter can also be inversed (5.16) as follows:

$$\Delta t_{\mathrm{rms}} = \frac{1}{2\pi f \times 10^{\mathrm{SNR}/20}} \tag{5.17}$$

Therefore, if we take a 10-bit ADC as an example, calculate SNR = 6.02 × 10 + 1.76 about 62 dB, which is a 200-MS/s ADC. Therefore, for Nyquist's requirements, if the input frequency f is 100 MHz, a 10-bit SNR requires a clock jitter of less than 1.26 ps RMS.

5.1.4.2 Thermal Noise Limit

The kT/C noise is generated by the unavoidable thermal noise associated with the sampling switch. It can be observed in all actual sampling data systems. When sampling, the storage capacitor C is connected to V_{in} through a CMOS switch. The on-resistance (R_{ON}) of the switch generates thermal noise, as presented in Fig. 5.9.

Obviously, for an infinite sampling capacitor or zero temperature, the kT/C noise will become zero. This is why it is described as a fundamental limitation of any practical sampling data system. To calculate noise, we need to take the input spectral density, multiply it by the output spectral density, and then integrate the total frequency, so the square of the spectral density is the square of the low-pass function. First, the output transfer function is expressed as follows:

$$V_{\mathrm{out}}(s) = V_{\mathrm{in}}(s) \frac{1}{1 + sRC} \tag{5.18}$$

The output noise spectral density is as follows:

$$S_{\mathrm{n}}(f) = S_{\mathrm{r}}(f) \left| \frac{1}{1 + j2\pi fRC} \right|^2 \tag{5.19}$$

The expected noise power is as follows:

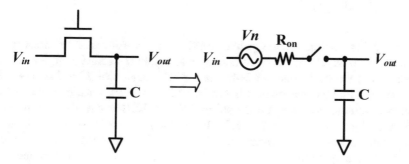

Fig. 5.9 Thermal noise generated by switch on-resistance

$$v_n^2 = 4kTR_{ON} \int_0^\infty \frac{1}{1+(2\pi fRC)^2}\,df = \frac{4kTR_{ON}}{2\pi R_{ON}C} \int_0^\infty \frac{1}{1+x^2}\,dx$$

$$= \frac{2kT}{\pi C} \int_0^{\frac{\pi}{2}} d\theta = \frac{kT}{C} \qquad\qquad (5.20)$$

The noise power result is independent of R_{ON}, a fundamental physical law. This is a bit intuitive because a smaller R_{ON} value will reduce thermal noise but increase noise bandwidth. When we calculate noise, noise magnitude and bandwidth have opposite effects on resistor, so only the capacitor can determine the noise. This is also a well-known thermal noise formula. The value of thermal noise in a typical CMOS switch resistor is 100 to 300 electron rms (about 60- to 180-mV rms). k is the Boltzmann constant ($= 1.38 \times 10^{-23}$ J/K).

5.1.4.3 Calculation Example of Noise Requirements

If the SAR-ADC is considered, the noise $\sqrt{kT/C}$ contributed by the switch only needs to be less than or equal to the quantization noise $\Delta/\sqrt{12}$. However, if we consider a pipelined ADC, we must also consider the noise contributed by the amplifier. Therefore, the design of the circuit can only limit half of the noise, so $\sqrt{kT/C}$ must be less than or equal to the quantization noise $\Delta/2\sqrt{12}$. If we want to design a 12-bit pipeline ADC, with a reference voltage of 1 V and sampling time of 200 MS/s, we need a bandwidth of about 1 GHz and to calculate half of the quantization noise. If the voltage is $\Delta/2\sqrt{12} = 35$ μV, we need to use a sampling capacitor of 3.3 pF or more. Because of the known noise voltage, according to $v_n^2 = 4\,kTBR_{ON}$, the RON can be calculated to be 75 ohms. From this, the size of the CMOS switch can be calculated.

5.2 Introduction to A/D Converter Architecture

5.2.1 Flash ADC

Flash ADC (sometimes called parallel ADC) is the fastest and most direct type of ADC, using a large number of comparators to compare against a large number of references at the same time. An N-bit flash ADC consists of 2^N resistors and 2^{N-1} comparators, and its arrangement is presented in Fig. 5.10. Each comparator has a reference voltage from the resistor string that is 1 LSB higher than the reference voltage of the next resistor in the chain. For a given input voltage, the input voltage of all comparators below a certain point will be greater than the reference voltage, and the logic output will be "1"; the reference voltage of all comparators above this point will be greater than the input voltage and output a logic "0." Similar to a

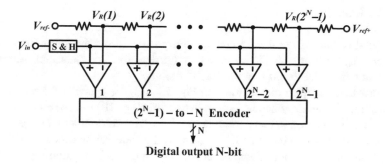

Fig. 5.10 Flash A/D converter

mercury thermometer, the output code at this time is sometimes called a thermometer code. Since the 2^{N-1} data output is impractical in practice, the decoder processes it to generate an N-bit binary output.

As can be seen from Fig. 5.10, the input signal is immediately added to all the comparators, so the output of the thermometer code is only delayed by one comparator delay than the input, and the N-bit output of the encoder is only delayed by a few gate delays; thus, this process is a bit complicated but very fast. This architecture uses a large number of resistors and comparators, so it is limited and only suitable for low resolution. If the speed is to be increased, each comparator must operate at a relatively high power level. Therefore, the problems of flash ADCs include limited resolution, high power consumption due to a large number of high-speed comparators, and relatively large chip size. In addition, the resistance of the reference resistor chain must be kept low in order to provide sufficient bias current for the fast comparator.

5.2.1.1 Clock Skew

Ideally, the comparators in the flash converter match greatly in terms of DC and AC characteristics. Since the sampling clock is added to all comparators at the same time, the flash converter is essentially a sampling converter. In fact, the delay is different among the comparators, and other AC mismatches can cause the ENOB to drop at high input frequencies. This is because the input conversion rate is equivalent to the conversion time of the comparator. Due to the inconsistent arrival times, routing the clock from comparator 1 to comparator 2^{N-1} will cause clock skew. This means that different comparators will sample the input at different time instances, resulting in poor performance at higher input frequencies. Therefore, it is often necessary to sample and hold before the flash converter. As can be seen from Fig. 5.10, sample and hold accomplish two things. First, it provides a hold signal to the comparator array, thereby minimizing the impact of clock deviation. Second, it provides good input impedance for driving any stage of the ADC and achieves high SFDR on high-frequency input signals.

5.2.1.2 Encoding

The digital output of the flash converter comparator forms the thermometer code. Usually, only one output is "1" through the edge detector. The metastable state of the comparator may also result in multiple "1" outputs.

As can be seen from the example in Fig. 5.11(a), the spark in the thermometer code changes the binary encoder output from b2b1b0 = 010 to b2b1b0 = 110. If the binary code is changed to that shown in Fig. 5.11(b) Sparkle-tolerant edge detector, isolated bubbles can be prevented. The correct edge detector detects the "010" edge. In addition, there are more complex coding schemes, such as counting "1" in the thermometer code.

5.2.2 Pipeline ADC

The principle of the pipeline A/D converter is based on a two-step recursive implementation of a more refined ADC. The pipeline A/D converter uses each stage of the cascade. In each stage, the input signal is sampled and held, and each stage performs the basic functions required by the timing algorithm. Each stage of the pipeline produces two outputs; a code represents the quantized value, and the difference or residue between the codes is subtracted from the input. And amplify the residue by a certain multiple, and then pass it to the next stage. The accuracy of the amplified analog signal must reach the number of bits to be determined at this stage. Because it can use K stages in parallel, the first input signal is sampled and

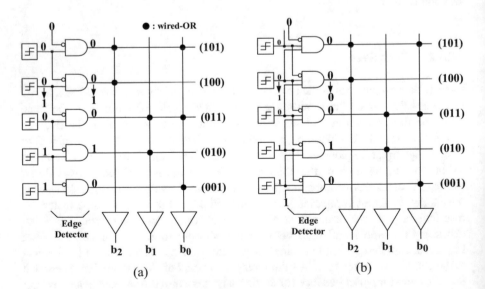

Fig. 5.11 Flash A/D encoder, (**a**) binary encoding, (**b**) sparkle-tolerant edge detector

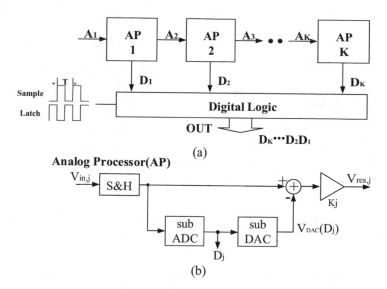

Fig. 5.12 (a) Pipeline ADC architecture, (b) single-stage analog processor

held N times before obtaining the final digital output and converted bit by bit after K clock delays.

Figure 5.12(a) presents the conceptual block diagram of a pipeline architecture with K stages. This timing scheme assumes that the duty cycle of the clock is 50% for best performance. One clock phase is used for sampling, and the other clock phase is used for latching the comparator. After latching, each sampling stage will produce an analog output for sampling in the next stage of the pipeline. The first stage generates D_1 bits, the second stage lowers the terminal D_2 bits, and so on. Therefore, the entire pipeline generates $D_1 + D_2 + \ldots + D_K$ bits. Digital logic combines the bits from each logic unit with a stage and generates an output word at a rate of f_s. This generates a so-called latency period of K clock cycles before the samples entering the pipeline are converted at the output. The time caused by this delay is the result of the pipeline operation. Since the K stages are used in parallel, the throughput of the pipelined ADC is increased by K times. In most of the advanced pipelines, each stage is implemented using switched capacitor technology. In view of the continuous expansion of technology, the latest pipeline also uses complex algorithms to correct the ADC, DAC, and amplification errors. The pipeline can also generate multiple bits at each stage. If this is the case, each stage requires a multi-bit ADC to obtain a digital output and a multi-bit DAC to generate the input of the next stage. The total resolution of the pipeline architecture comes from the sum of the bits in each stage. Please note that the number of bits in each level can be equal to or different from each other according to the design trade-offs.

The block diagram of the general pipeline-level analog processor is presented in Fig. 5.12(b). The ADC generates j bits, and the DAC uses the same number of bits to convert the result to analog. In fact, the DAC resolution of the architecture using

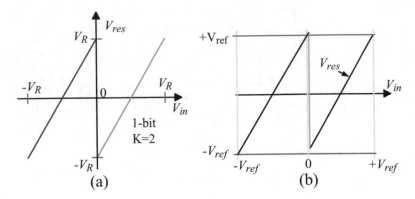

Fig. 5.13 (**a**) Transfer characteristics of a 1-bit residue generator, (**b**) residual response with real ADC and ideal DAC

digital correction is lower than that of the ADC. The V_{in} quantization error obtained by subtracting the output of the DAC from V_{in} determines the new residual voltage after amplification:

$$V_{res,j} = \left[V_{res,j-1} - V_{DAC}(D_j) \right] \cdot K_j \qquad (5.21)$$

If the DAC gain of the D_j bit is 2^{Dj}, the dynamic range of the residual is equal to the dynamic range of the input.

Figure 5.13(a) presents the ideal input and output transfer characteristics of the residual generator. Its input range is $-V_R \sim +V_R$, 1-bit ADC and DAC, $K = 2$. Under negative input ($-V_R \sim 0$), the D_j output is 0, please refer to the red line in the figure. After VDAC(0) $= -VR/2$ multiplied by 2, the remaining voltage range is equal to $-VR \sim + VR$. At $- VR/2$, the residual is greater than 0, as is the positive input ($0 \sim +$ VR), refer to the green line in the figure. The output of D_j is 1, $V_{DAC}(1) = V_R/2$ multiplied by 2, and the voltage range is also $-V_R \sim + V_R$.

Because the remaining number of bits to be estimated is decreasing throughout the pipeline, the accuracy requirements and design difficulties of the first few stages of the pipeline are greater than those of the last stage. The most demanding requirement is the input S&H requirement, because the imperfection of the first module causes the offset, gain error, and nonlinearity of the entire pipeline.

The performance of pipelined ADCs is limited by the non-idealities of ADCs, DACs, and inter-stage amplifiers used in the pipeline stages. The threshold error in the ADC causes the residual amplitude to be greater or less than the full scale at the breakpoint, as presented in Fig. 5.13(b). The figure presents the signal produced when an actual ADC is used with an ideal DAC and an inter-stage amplifier. Residual responses highlight local anomalies because they can still correctly provide the difference between the analog input and the quantized signal generated by the DAC. Therefore, no error will occur before the next stage of the ADC, because it will not be able to correctly convert the remaining signal to a working range beyond $\pm V_{\text{ref}}$.

By using digital correction technology, errors caused by the ADC defects can be eliminated. It was previously observed that the dynamic range of the residue may exceed the expected limit. One method is to add redundant sub-range levels to the ADC at this stage. Increasing these redundancy levels can prevent the generation of out-of-range residuals, on the one hand, and prevent the provision of information to the digital domain, on the other hand; thus, ADC errors can be completely compensated for [3].

5.2.3   Successive Approximation ADC (SAR ADC)

The successive approximation ADC is essentially a DAC plus a comparator, as presented in Fig. 5.14. The idea here is that if we have a DAC, it is a high-speed DAC with an order of magnitude accuracy of the converter itself. Moreover, possible sub-regions are searched in binary to reduce the search time. After passing through the sample-and-hold circuit, the input voltage is compared with the DAC output voltage. Starting from the MSB, the fraction corresponding to each bit of the V_{FS} is sequentially added to the fraction corresponding to the determined bit, and the sum V_{da} is compared with the input sampling voltage. The entire program performs N comparisons. Until the error voltage is as small as possible, that is, the highest accuracy of the system is reached, the code represents the digital approximation of the input signal.

The detailed timing is described as follows. In the first stage, the input signal is accurately sampled. After this phase, the input signal will remain unchanged. Then the binary search method, the fastest search algorithm, is applied. We set all bits to zero; only the first bit is set to 1, $V_{\text{da}} = 1/2\ V_{\text{FS}}$. Because the input signal is higher

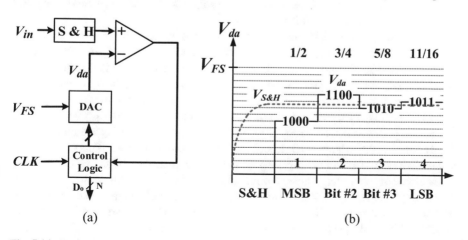

(a)   (b)

Fig. 5.14 Basic circuit diagram of the successive approximation algorithm

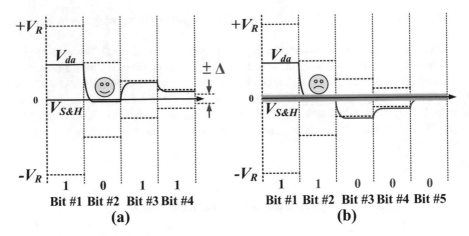

Fig. 5.15 (a) Correct search path, (b) search path with error at the second clock period [4]

than the V_{da}, always set the first bit to 1. Then, we move to the second bit and set it to 1, $V_{da} = (1/2 + 1/4) \cdot V_{FS}$. The input signal is compared with V_{da}. We can see that the input signal is lower than the V_{da}. In this case, the second bit must be set to zero. Subsequently, the third bit is set to 1, $V_{da} = (1/2 + 1/8) \, V_{FS}$. The input signal is compared with the V_{da}. We can see that the input signal is higher than the V_{da}, and the third bit is still set to 1. Then, we turn to the last bit and set it to 1, $V_{da} = (1/2 + 1/8 + 1/16) \cdot V_{FS}$. Once again, we can see that the input signal is lower than the reference voltage; thus, the last bit must be set to zero. At the end of this operation, we get the following output values: 1, 0, 1, 1.

The above example uses one clock cycle for S&H, and each subsequent clock cycle is used to determine each bit, so the N-bit conversion requires $(n + 1)$ clock intervals. If the S&H time of one cycle is not stable enough, it is sometimes convenient to use two clock cycles for sampling, so the conversion requires $(n + 2)$ clock intervals in total.

As presented in Fig. 5.15, any estimation error in the bit will propagate the search path along all successive steps in some way, thereby modifying the search path. Suppose that V_{da} changes from a level much higher than $V_{S\&H}$ to a slightly lower level (the second clock cycle in the figure). It may happen that the comparator does not recover from the overdrive state fast enough, and eventually the comparator determines a logic 1 instead of a logic 0. This will cause the next V_{da} voltage to lead the search path toward the wrong direction and follow the resulting search path. The final code is 1100 instead of 1011, and the error starts at position 2. This error usually occurs when a large switching step in the search path causes an overload in the comparator or due to large noise from analog switches and comparators. Redundant comparison is the way to correct this error, that is, the search range must be extended at the end to accommodate the initial error. However, this extended search will require additional clock cycles to complete the algorithm.

Redundant comparison can tolerate noise within $\pm\Delta$. The required correction value is determined as follows according to the results of Bit#4 and Bit#5:

$$D_e = \begin{cases} +1 \text{ if Bit\#4} = \text{Bit\#5} = 1 \\ -1 \text{ if Bit\#4} = \text{Bit\#5} = 0 \\ 0 \text{ if Bit\#4} \neq \text{Bit\#5} > 0 \end{cases} \qquad (5.22)$$

Because Bit#4 = and Bit#5 = 0, the last code in the above example is 1100, and we need to subtract 1 to get 1011.

5.2.4   Sigma-Delta Data Converter

The advantages of sigma-delta data converters include low sensitivity to circuit defects, such as drift and temperature, inherent monotonicity and linearity, and almost no DNL. This is a good compromise between speed and resolution and digital complexity. Therefore, it is used in many applications that require low-cost, low-bandwidth, low-power, and high-resolution ADCs. The disadvantage is that there is a long delay between the beginning of the sampling period and the first valid digital output, because oversampling and noise shaping require filtering.

The operation of the Σ-Δ ADC is not particularly difficult to understand. Simply put, the Σ-Δ modulator is oversampling + noise shaping. By shaping the energy spectrum of the quantization noise, most of the noise appears outside the bandwidth of interest, thereby improving the low-frequency SNR. The digital filter then removes the noise outside the bandwidth of interest, and the decimator reduces the output data rate back to the Nyquist rate.

We first discussed sampling techniques, which are easier to understand through frequency domain analysis. The data sampling system has quantization noise. There is high quantization error energy near the DC in the frequency domain. The RMS quantization noise $\Delta/\sqrt{12}$ of a perfect classic N-bit sampling ADC is uniformly distributed in the Nyquist band from DC to $f_s/2$ (where Δ denotes the LSB, and f_s denotes the sampling rate), as presented in Fig. 5.16(a). Therefore, the SNR of its full-scale sine wave input will be $(6.02 N + 1.76)$ dB. If the ADC is not perfect and its noise is greater than its theoretical minimum quantization noise, its effective resolution will be less than N bits. The actual resolution will be defined as: ENOB = $(\text{SNR} - 1.76)/6.02$.

If we choose a higher sampling rate $K \cdot f_s$ (Fig. 5.16(b)), the RMS quantization noise still exists $\Delta/\sqrt{12}$, but the noise is now distributed over a wider bandwidth from dc to $K \cdot f_s/2$. Then, if a digital LPF is applied to the output, a lot of quantization noise will be removed, but the desired signal will not be affected; thus, the ENOB is improved. We have completed the high-resolution A/D conversion with a low-resolution ADC. The factor K is usually called the oversampling rate (OSR = $f_s/2 \cdot fB$). If we simply apply oversampling to increase the resolution

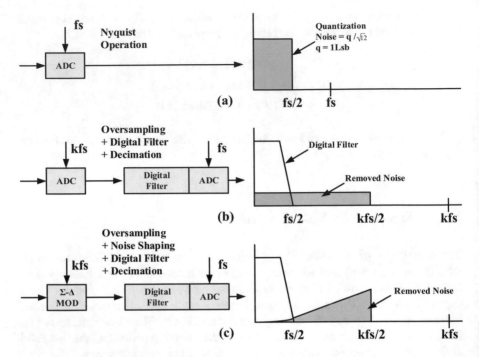

Fig. 5.16 Oversampling, digital filtering, noise shaping, and decimation

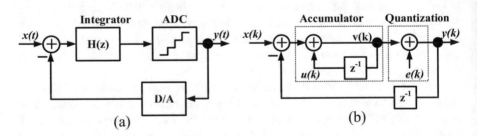

Fig. 5.17 First-order sigma-delta ADC

ENOB = N + 0.5 log$_2$ (OSR), we must perform oversampling by a factor of 2^N to obtain an N-bit resolution increase. For example, to obtain 5 bits, OSR = 1024 must be used, which is not very cost-effective. However, the sigma-delta converter does not need such a high oversampling rate, because not only is the signal passband limited, but the shape of the quantization noise makes most of it fall outside the passband, as presented in Fig. 5.16(c) [5].

Next, we will introduce noise shaping. In Fig. 5.17(a), the first-order sigma-delta ADC contains two basic modules: an integrator and a 1-bit ADC (usually called a comparator). The integrator and comparator can be modeled as accumulator and quantizer, respectively, as presented in Fig. 5.17(b), where $e(k)$ is the quantization noise generated by the quantization process. For an ideal quantizer with a step size of

Δ, it is assumed that the pdf of $e(k)$ is uniformly distributed on $-\Delta/2$ and $+\Delta/2$. The transfer function in Fig. 5.17(b) can be written as follows:

$$y(k) = x(k) + e(k) - e(k-1) \tag{5.23}$$

In z-domain,

$$Y(z) = X(z) + \left(1 - z^{-1}\right) E(z) = \text{STF}(z)X(z) + \text{NTF}(z)E(z) \tag{5.24}$$

$$\text{Signal transer function STF}(z) = \frac{Y(z)}{X(z)} = 1 \tag{5.25}$$

$$\text{Noise transer function } NTF(z) = \frac{Y(z)}{E(z)} = 1 - z^{-1} \tag{5.26}$$

The NTF in the frequency domain is

$$NTF(f) - \text{NTF}(z)|_{z=e^{j2\pi f/f_s}} = \sin\left(\frac{\pi f}{f_s}\right) \times 2j \times e^{j\pi f/f_s} \tag{5.27}$$

The quantization noise power of the f_B band is

$$P_n = \int_{-f_B}^{+f_B} S_e(f)|\text{NTF}(f)|^2 df = \int_{-f_B}^{+f_B} \frac{\Delta^2}{12 f_s}\left[2\sin\left(\frac{\pi f}{f_s}\right)\right]^2 df$$

$$\approx \frac{\Delta^2}{12} \cdot \frac{\pi^2}{3} \cdot \left(\frac{2 f_B}{f_s}\right)^3 = \frac{\Delta^2}{12} \cdot \frac{\pi^2}{3} \cdot \frac{1}{\text{OSR}^3} \quad \text{if OSR} \gg 1 \tag{5.28}$$

$$\text{SNR} = \frac{12}{8} \cdot \frac{3}{\pi^2} \cdot \frac{1}{\Delta^2} \text{OSR}^3 \tag{5.29}$$

$$\text{SNR} = 6.02N - 3.41 + 9.03 \cdot \log_2(\text{OSR}) \tag{5.30}$$

NTF can suppress noise or harmonics caused by the nonlinear characteristics of the integrator and quantizer to improve the SNR. After noise shaping and oversampling, the SNR is improved by 9 dB/octave or 1.5 bit/octave.

Due to the existence of 1-bit ADC and 1-bit DAC, the simple first-order unit sigma-delta ADC is inherently linear and monotonic, but it cannot provide sufficient noise shaping for high-resolution applications. Increasing the number of integrators in the modulator (similar to the addition of poles to the filter) can provide more noise shaping, but at the cost of a more complex design: the second-order 1-bit modulator presented in Fig. 5.18 [5]. Although higher-order modulators improve the noise shaping characteristics, those greater than third order are difficult to stabilize and pose significant design challenges.

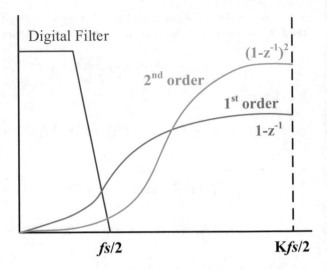

Fig. 5.18 Noise shaping characteristics of first- and second-order modulators

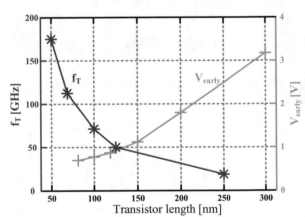

Fig. 5.19 Device speed (f_T) in the CMOS technology trend

5.3   SAR Analog-to-Digital Converter

5.3.1   Advantages of SAR ADC

In the scaling CMOS technology trend presented in Fig. 5.19 [6], there are two very obvious trends in gate length reduction. First, the operating voltage and operational amplifier gain are decreasing. Second, the speed of the device (f_T) is increasing. SAR A/D converters can take full advantage of this trend, because the topology of SAR A/D converters avoids the use of operational amplifiers and can use very fast components in the latest technology. In addition, the small area and ultra-LP consumption of SAR itself make it more advantageous for the applications, and the use of interleaved architecture can also allow SAR to improve in higher-speed applications. When we want to check the quality factor to show power divided by speed, SAR is the best A/D converter.

5.3.2 SAR Algorithm

The functional block of the N-bit SAR ADC is presented in Fig. 5.20. It includes an S&H, a DAC, a comparator, and a SAR controller. We only define 1 bit in any one clock cycle, and it takes N clock cycles to execute the N-bit digitization process, plus one for input sampling. Therefore, the conversion time is approximately N + 1 times the clock period. This means that for a given data rate (f_s), the external data rate, the internal circuit runs at $(N + 1) \cdot f_s$.

Since there is no gain stage or amplifier, SAR-ADC will not be affected by the scaling technology and cause performance degradation. It is precisely because each stage of conversion does not introduce any gain, unlike pipeline ADCs; the accuracy will be reduced after each conversion is completed. Therefore, each block must run with the target accuracy within the available time slot, and all errors are equally important. Since there is no operational amplifier, the power consumption is very low and is only consumed by the comparator, DAC, and logic circuit, which are all-digital systems with only dynamic power consumption.

5.3.2.1 SAR-ADC Accuracy Limits

Both S&H and DAC will affect the accuracy performance of SAR-ADC. Because any errors generated in the S&H path cannot be recovered by the process itself, a high-precision and very accurate sample-and-hold circuit is required. In addition to avoiding noise from the switch itself, it is also important to avoid switch charge injection and clock feedthrough. The capacitor needs stable charging and discharging times, low clock jitter, and accurate setting time.

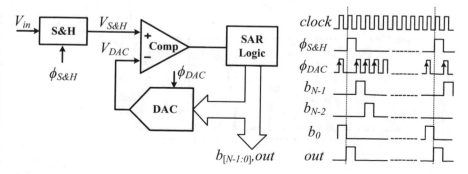

Fig. 5.20 Functional blocks of the N-bit SAR ADC

The accuracy of the DAC also determines the accuracy of the A/D converter. The DNL and INL of the DAC will become the DNL and INL of the A/D converter. In a typical integrated circuit, the DAC is usually implemented by an integrating unit array. The accuracy of SAR-ADC can reach 9 to 11 bits without any other operation calibration adjustments. Since the DAC and related accuracy are limited by an individual component mismatch, we can also achieve accuracy beyond technical mismatch through factory fine-tuning and on-chip calibration.

The comparator has different restrictions compared with the sample-and-hold circuit and DAC. There are two possible errors of the comparator that will affect the judgment of the comparator. The first error is the offset, that is, the DC voltage. The offset can be directly modeled and placed on the input of the comparator, which appears as an offset in the overall transmission characteristics. It will not affect the overall linearity of the ADC. The offset of the comparator can be eliminated by applying offset elimination (auto-zero) technology. The second error comes from noise. Noise can produce random errors, just like randomly changing the threshold voltage of a comparator, which can lead to judgment errors. Therefore, the noise must be reduced below the LSB accuracy.

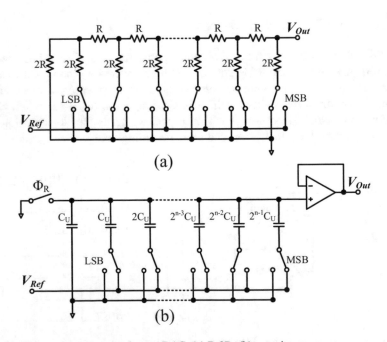

Fig. 5.21 Two main ways to implement DAC: (**a**) R-2R, (**b**) capacitor array

5.3.2.2   DAC Implementation Technology

There are two main ways to implement DAC. One is to use the resistor presented in Fig. 5.21(a), and the other is to use the capacitor presented in Fig. 5.21(b). They have different functions. The biggest problem with resistors is that they have static power consumption; thus, they cannot be used in LP applications. Another problem is that the resistor cannot sample the input signal and requires additional S&H. In the capacitive DAC, because the capacitor has a memory function, it can combine two different functions of sampling and holding at the same time to reduce the number of components and area. Capacitive DACs have no static power consumption, so power consumption is only dynamic. In the general manufacturing process, capacitance matching is better than resistance matching. But in some high-resolution DACs, it can be observed that the DAC uses part of the resistance and part of the capacitance, because the resistance can be modified by fine-tuning the resistance to achieve matching to obtain high accuracy; the capacitance does not have this mechanism.

5.3.2.3   DAC Encoding

On the DAC, different resistor or capacitor topological arrangements correspond to different codes. The simplest DAC code is to use the thermometer code because it uses the same value; it is very monotonous, but the wiring is very complicated. Considering the wiring and DAC aspects, the typical method is usually a binary DAC. If we want to achieve good monotonicity, we must ensure that the mismatch is less than ½ LSB, which corresponds to a larger unit component. Therefore, a large area is required to realize the monotonic DAC. We can consider hybrid or segmented methods: some use binary, some use binary in combination with thermometers, and some use redundant coding techniques to improve the performance of the DAC by using a lower index than binary (for example, 1.8). Certainly, these larger areas corresponding to binary DACs include additional algorithms and logic control, requiring longer conversion times, but mismatches and corrections can be made.

Regarding the implementation example of the resistance DAC, paper [7] in Fig. 5.22 uses the R-2R resistance DAC embedded in the module to achieve 14-bit high precision. As aforementioned, resistive DAC consumes a lot of power, and because the input cannot be sampled, it requires external active S&H and auto-zero operation to eliminate offset and store redundant calibration data in EEPROM. This is the main method for traditional mass production of ADCs because it is very stable and can increase the DAC resolution beyond the matching range. However, laser trimming is required at the factory level, and sometimes heat is very sensitive to the mechanical stress of the package. This operation requires additional time and expense. Recent scaling techniques allow the introduction of very effective calibration algorithms, namely, on-chip digital calibration algorithms. Although the calibration algorithm is complicated, the area is reduced, and the power consumption is extremely low.

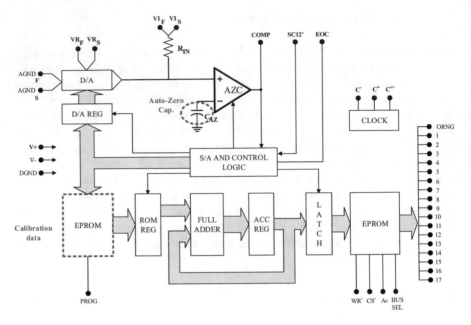

Fig. 5.22   14-bit CMOS A/D converter with error correction [7]

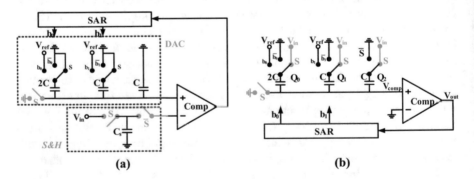

Fig. 5.23   Capacitive DAC implementation

Here, another DAC that uses capacitors is introduced. The realization of capacitive DAC charge distribution is presented in Fig. 5.23. In Fig. 5.23(a), there are independent S&H and DAC. Furthermore, there is a capacitive DAC controlled by a 2-bit DAC and sample-and-hold circuit. At the top of the figure, the result of the DAC is connected to the input of the comparator. At the bottom of the figure, sample and hold is applied to the other input of the comparator. The sampling capacitor Cs is the load of the previous circuit, and the DAC capacitor is the load of V_{ref}. The sample-and-hold circuit design should minimize the thermal noise of sampling and the charge injection clock feedthrough of the switch. The DAC must consider thermal noise and process capability to ensure target accuracy. Figure 5.23(b) presents a solution that combines two different functions. The sample-and-hold capacitor and the DAC capacitor share the same capacitor array.

Figure 5.23(b) presents the structure of a 2-bit A/D converter. It works as follows. During the sampling phase, all the bottom plates of the capacitor array are connected to the V_{in}, the top plate is grounded, and the full array capacitors are precharged to V_{in}. After sampling, the charge on the entire array is $Q_{tot} = Q_0 + Q_1 + Q_2 = 4CV_{in}$. In the next conversion phase, the top plate of the capacitor array remains open in a floating state, and Q_{tot} (i.e., the sampling input signal) remains unchanged until the end of the conversion phase.

In the subsequent conversion phase, the smallest capacitor b2 must always remain grounded. Start from the MSB, and then guess the MSB ($b0 = 1$); the bottom plate of the largest capacitance is connected to V_{ref}, and the other part of the array is grounded. There is no additional charge injection during this clock phase. Then, after the DAC stabilizes quickly, the equation is established according to the law of conservation of charge, and the voltage on the top plate is expressed as follows:

$$-4CV_{in} = Q_{tot}(\text{Sample}) = Q_{tot}(\text{Convert}) = 2C(V_{cmp} - V_{ref}) + 2CV_{cmp} \quad (5.31)$$

$$V_{cmp} = V_{ref}/2 - V_{in} \quad (5.32)$$

The comparator input is the difference between $V_{ref}/2$ and V_{in}. This determines the value of MSB (b0) and fixes it for the following operations. The first number describes the $V_{ref}/2$ voltage, the second number describes the $V_{ref}/4$ voltage, and so on. The SAR algorithm is summarized as follows:

SAR algorithm:

j from 1 to N and judges 1 bit each time:

On each transition stage, b_j is guessed to be 1, the comparison is complete, and V_{out} is generated

- If $V_{out} = 1 \rightarrow$ guess OK, then use $b_j = 1$ in the next step.
- If $V_{out} = 0 \rightarrow$ guess wrong, then use $b_j = 0$ in the next step.

Then, select the second bit, third bit, and so on, until the last one.

The input of the comparator is the binary representation of V_{ref} minus the input voltage:

$$V_{cmp} = \left(\sum_{j=1}^{N} \frac{V_{ref}}{2^j} \cdot b_j \right) - V_{in} \quad (5.33)$$

5.3.2.4 Conversion Time

The number of digits to be analyzed is exactly equal to the number of conversions in the sequence. Then, at the end of the conversion phase, it enters the sampling phase, where the input signal is extracted. Therefore, the complete conversion phase is not only the number of conversion clock cycles but also the sampling of the input signal. Usually, sampling is not limited to one clock cycle, but sometimes depending on the accuracy of the A/D converter we want, it can exceed one clock cycle (usually two or three clock cycles), which also increases the conversion time.

$$t_{conversion} = N \cdot t_{clk} + N_{samp} \cdot t_{clk} = (N+1) \cdot t_{clk} = 1/f_s \qquad (5.34)$$

In Fig. 5.23(b) Binary capacitor DAC structure, the minimum unit capacitance C is selected based on thermal noise and capacitance matching. The unit capacitance must be large enough, and the thermal noise must be low enough to ensure the target SNR; moreover, the capacitance matching must be better than the DAC requirements. Capacitance mismatch is usually caused by process capability limitations and excessive parasitic capacitance. Considering the process and parasitic capacitance, the unit capacitance must be large enough. For example, in the case of a 10-bit A/D converter, if the value of the unit capacitor is 50 fF, since there are 2^{10} unit capacitors in our array, the total capacitance of the capacitors is 50 pF, which is a large load for the previous stage and imposes a speed limit on S&H.

5.3.2.5 Switch Resistance Requirements

In the entire A/D conversion process, any error caused by any step (sample hold and conversion) will make the ADC result wrong. Therefore, sampling and holding must be very accurate. When performing passive sampling, to obtain high accuracy, due to the time constant, a very long window must be used, and it will increase the conversion time of the A/D converter. The use of parallel sampling or boot switches can be considered to avoid the increase in conversion time. The speed of the SAR-ADC is also limited by the switching speed of the DAC. Because the size of the capacitor array is proportional to 2^N, the capacitor array must be charged and discharged within one clock cycle. The conversion rate during sampling and

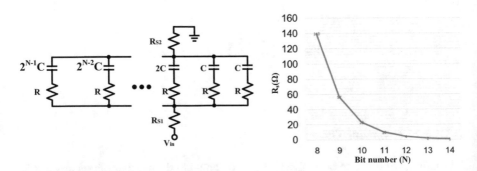

Fig. 5.24 Relationship between switch resistance and the number of bits

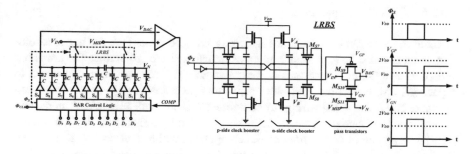

Fig. 5.25 SAR-ADC architecture with the LRBS schematic and waveforms of the original and boosted clock

conversion is limited by the time constant of the RC network, as presented in Fig. 5.24. A rough estimate of the time constant is as follows:

$$\tau_{eq} = (R_{s1} + R + R_{s1}) \cdot 2^N \cdot C \approx R_s \cdot 2^{N+1} \cdot C \qquad (5.35)$$

Must be stable within 1/2LSB, i.e.

$$e^{-T/\tau_{eq}} < \frac{1}{2^{N+1}} \qquad (5.36)$$

$$t_{clk} > \tau_{eq} \cdot (N+1) \cdot \ln(2) = 0.69\tau_{eq} \cdot (N+1) \qquad (5.37)$$

$$\tau_{eq} = R_s \cdot 2^{N+1} \cdot C < \frac{t_{clk}}{0.69 \cdot (N+1)} = \frac{1}{f_s \cdot 0.69 \cdot (N+1)^2} \qquad (5.38)$$

$$R_s < \frac{1}{f_s \cdot 2^{N+1} \cdot C \cdot 0.69 \cdot (N+1)^2} \qquad (5.39)$$

It must be ensured that all these networks stabilize to the final resolution within a given clock cycle. If $f_s = 5$ MHz, $C = 50$ fF, and the number of bits is 8 to 14, the required switch resistance R_s is presented in Fig. 5.24.

In order for the CMOS switch to have such a small resistance, a larger switch size is required. However, large CMOS switches have problems of charge injection, clock feedthrough, and leakage. In Fig. 5.25 [8], the bootstrap switch is used to solve the problem of input S&H and speed limit. They use a low-leakage bootstrap switch (LRBS) to achieve a satisfactory ENOB. In order to reduce the switch resistance R_{ON}, they increase $I_{DS} \propto (V_{GS} - V_T)^2$ and pump V_{GS} to nearly twice the value of V_{DD}. In order to increase the switch resistance R_{OFF}, that is, to reduce the subthreshold leakage current $I_{sub} \propto \exp.[q(V_{GS} - V_T)/nkT]$, the V_{GS} of the transistor is reduced to a negative voltage.

5.3.2.6 Parasitic Capacitance

In addition to the DAC capacitance matching must be better than the target resolution. It should also be noted that any parasitic capacitance will reduce the matching and ADC resolution. These parasitic capacitances include windings connected to capacitors, unit capacitors, and other capacitors, or parasitic capacitances between the upper and lower metal wirings or components. Using a metal shield or increasing the distance to reduce parasitic capacitance may also be considered. In addition, the parasitic capacitance at the input of the comparator will also affect the accuracy of the ADC. In advanced manufacturing processes, dummy metal filling is used to obtain a uniform etch profile through a CMP (chemical mechanical polishing) process and maintain overall chip-level flatness. However, this floating metal filling will more or less affect the parasitic capacitance. The use of these dummy metal fillers should be avoided when implementing DAC capacitors, because this will cause the accuracy and operating speed of the ADC to deviate from the designer's expectations.

5.3.3 Comparator Offset Calibration Technology

Comparator offset calibration techniques can be divided into two basic categories: background and foreground. In the background calibration, the calibration is completed during normal operation, and the calibration value needs to be determined periodically. In the foreground calibration, the required calibration value is determined once during system startup or just before the sub-block operation, and the obtained value is stored and used to correct the offset during normal operation.

5.3.3.1 Background Calibration Technology

Background calibration techniques can be used to correct static and dynamic offsets in comparators. Because additional clock cycles are required to determine the correction word for each comparison cycle, it is only used in low-speed applications. The background calibration technique widely used in SAR-ADCs is the auto-zero technique.

In the auto-zero technique, the comparator offset is sampled by a switched capacitor on the input or output side of the comparator and subtracted from the signal before the next clock cycle to eliminate the offset [9]. One disadvantage of this technique is that it requires multiple switched capacitors, which limits the sampling rate and bandwidth. In addition, it can only be used with a clock comparator. The offset sampling technique introduces additional capacitance in the signal path. They also cannot completely eliminate the offset voltage of the latching comparator because the elimination is limited by the mismatch between the MOS switches used by the charge injection or clock feedthrough in the correction circuit.

5.3.3.2 Foreground Calibration Technology

The foreground calibration technique can only correct static offsets. Performing foreground calibration is not as complicated as performing background calibration. Therefore, it is more suitable for high-speed comparators and ADCs. Common foreground calibration techniques include threshold adjustment and digital control fine-tuning methods.

Non-volatile floating gate devices can be used to adjust the threshold voltage of comparator transistors in critical signal paths. These components store the charge used for fine-tuning to correct the offset voltage [10]. Special process steps are required to construct stacked floating gates to realize this calibration technique.

Digital control fine-tuning is a popular calibration technique developed for high-speed ADCs, and it can be implemented in many ways. The difference between these calibration techniques and traditional methods is that the offset voltage is not corrected after sampling. Instead, the calibration circuit compensates for the input reference offset by applying a digitally controlled imbalance on one of the comparator's signal paths. Numerical control fine-tuning methods are divided into two methods: digital calibration and digital-assisted analog calibration.

The fine-tuning DAC balances the correction resolution by fine-tuning the step size, calibration time, power consumption, and silicon area. If a high-resolution DAC is used to set the analog fine-tuning step size smaller to achieving fine correction resolution, the calibration time will be longer, and the correction circuit will consume more power. In addition, a high-resolution integrated fine-tuning DAC requires a larger silicon area. Conversely, if a low-resolution DAC is used and the correction step size is larger, the correction time will be shorter, and the power consumed by the correction circuit will be smaller, as well as the silicon area. However, the resulting offset voltage will be relatively high.

Digital calibration is achieved by using digitally controlled binary weighting circuit elements (capacitors, transistors, etc.) connected to different comparator nodes [11, 12]. Conversely, digital-assisted analog calibration is achieved by applying analog calibration signals to offset the unbalanced component parameters (current, capacitance, transconductance, resistance, etc.) on the comparator signal path [13–15]. These calibration signals are generated by an integrated DAC, adding a single-circuit element to the comparator instead of the element array. First, through integrated or external digital logic or signal processing circuit, the digital correction word is determined by the statistics or linear search output by the equalization comparator. The resulting correction word is then stored in the integrated memory. The stored digital correction word either controls the value of the circuit element, such as a binary-weighted capacitive load, or controls the DAC that generates the analog trimming signal for the added circuit element. DAC calibration is also a trade-off between calibration resolution and fine-tuning step size, calibration time, power consumption, and silicon area.

5.3.3.3   Circuit Implementation of Foreground Calibration Technology

Three circuit technologies are widely used to implement digital control foreground offset calibration in comparators, namely, body voltage, shunt current, and capacitive load fine-tuning, which are presented in Fig. 5.26. Fine adjustments should be made according to the signal type. The first two are analog technologies, and the last one is digital technology.

The bulk voltage trimming technology is based on the balance of the comparator differential input stage (ID_p, ID_n) by applying analog trimming voltages (B1, B2) according to the polarity of the offset voltage to fine-tune the threshold voltages of the two input transistors, namely, M2 and M3, as presented in Fig. 5.26(a).

In the bulk voltage fine-tuning technology, since no additional capacitive load is introduced to connect to the signal path, this technology is more suitable for high-speed comparators [17, 18]. The main disadvantage of this technique is that it requires input transistors with isolation blocks. Therefore, this method can only be applied to comparators with PMOS input transistors or when using triple-well CMOS technology. Furthermore, this technology requires an additional power supply to provide a fine-tuning DAC for body voltage that is larger than the circuit power-supply voltage (V_{DD}).

Figure 5.27 presents an example of calibration using parallel current comparator offset voltage cancellation [17]. The architecture consists of a comparator, offset compensation current source (Mc1, Mc2), and a charge-pump. During the calibration mode, all the input nodes of the comparator are switched from the signal input to the common-mode voltage V_{cm}. The gate of Mc1 is connected to V_b to set the common-mode voltage of the charge-pump. The gate of Mc2 is connected to the capacitor C_H, which is precharged to V_b in the initial state. If the comparator has an offset voltage V_{offset} (V_{offset} is a positive value at this time), the comparator outputs a high level and draws the charge of C_H according to the current I_{cp}. As the control voltage of the current source V_c drops, the offset voltage of the comparator approaches zero. When V_c exceeds V_{offset} and the offset of the corresponding comparator is adjusted to zero, the comparator alternately outputs high and low

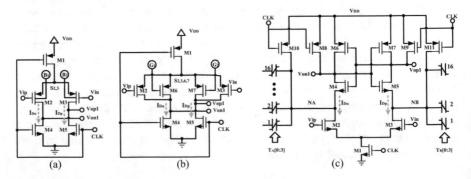

Fig. 5.26 Circuit techniques to implement foreground offset calibration in comparators; (**a**) bulk voltage, (**b**) shunt current, (**c**) capacitive load [16]

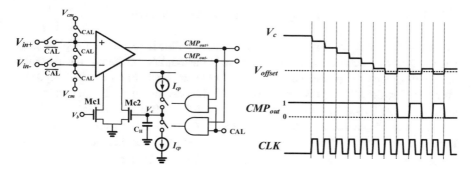

Fig. 5.27 Calibration architecture for offset cancellation

levels, as indicated by the waveform in Fig. 5.27. In the conversion mode, CH keeps the offset value. Therefore, the offset voltage is eliminated in the conversion mode.

5.3.3.4   SAR ADC Power Consumption

The power consumption of SAR-ADC is divided into two stages. First is the sampling phase. In this stage, the capacitor array is precharged by V_{in}; the capacitor array is the load of the previous circuit. Turning off other circuits (SAR, DAC, comparator, logic) can reduce the power consumption of the sampling phase. Second is the conversion phase. In this phase, the capacitor array is driven by V_{ref}, which accounts for about 80% of the power consumption. Other modules, such as comparator and SAR control logic, only consume dynamic power, which accounts for about 20% of power consumption. However, if a bootstrap switch is used, because it involves the sampling and conversion phases, it must be kept working all the time and the power consumption is directly related to the operating frequency. The SAR-ADC does not have a bias current, nor does it use an operational amplifier like a pipeline. Therefore, SAR power consumption is usually proportional to the operating frequency.

5.3.4   Trend of the Latest Development

The latest development has three main trends: low power consumption, high resolution, and high sampling frequency. Power consumption and resolution are related to the size of the capacitor, and the size of the capacitor is related to the sampling frequency, so these trends are interrelated.

5.3.4.1 Low Power Consumption–Dual Array Structure

For an N-bit capacitor array, 2^N times the unit capacitor is required. Since the large capacitor array leads to high driving power consumption, reducing the capacitor array can reduce the power consumption. One solution is to use a split capacitor array DAC structure, as presented in Fig. 5.28. Its structure aims to separate the two capacitor array DACs by attenuating capacitors (C_a). Each array resolves N/2 bits, so the maximum capacitance of each array is $2^{N/2-1}C$. The total capacitance is reduced from $2^N C$ to $2 \cdot 2^{N/2}C$. Since the sequence of C_a and the entire right array must be equal to C, the C_a series can be solved ($2^{N/2}C = C$) to obtain $C_a = \frac{2^{N/2}}{2^{N/2}-1} \cdot C$. Considering power consumption and process matching, the size of the unit capacitor should be as small as possible. However, the value of C_a is only a small part of the unit capacitance. Therefore, whether C_a can be accurately realized is the limitation of integrating various technologies.

5.3.4.2 Low Power Consumption–CAP Array Drive Strategy

The technique here to reduce the power consumption is the capacitor array drive strategy. When the DAC voltage changes from 1/2 V_{ref} to 3/4 V_{ref}, the energy consumption is $1/4C\ V_{ref}{}^2$. When the DAC voltage changes from 1/2 to 3/4 V_{ref}, switching b0 to GND and b1 to V_{ref}, the energy consumption is $5/4C\ V_{ref}{}^2$, because the capacitor 2C must be discharged, whereas the capacitor C must be charged. This is a double job. Figure 5.29 is to use the thermometer code to drive the capacitor, because the capacitor is only half discharged instead of discharging and charging, just switching the difference. In addition, another advantage of driving capacitors with thermometer codes is their linearity. But when connecting all the capacitors, the thermometer array is very complicated.

Recently, various capacitor switching schemes have been proposed to reduce the power consumption of SAR-ADCs. Energy is reduced using split capacitor [18], monotonic switching [19], and three-level switching methods [20]. Various schemes show that compared with traditional architecture, the average energy consumption is reduced by more than 37–98% [21]. By using top-plate sampling and V_{cm}-based

Fig. 5.28 Split capacitor array DAC structure

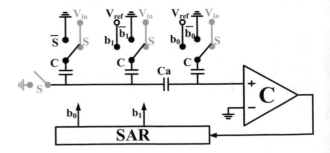

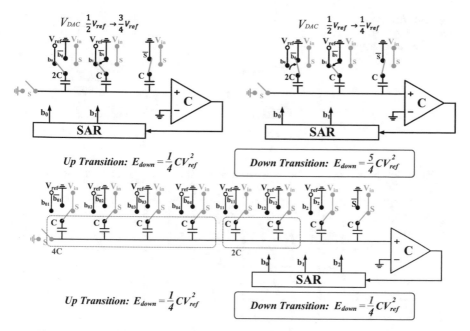

Fig. 5.29 Low-power SAR-ADC capacitor array driving strategy

switches, the number of capacitors in this solution is also reduced by 75% compared with that when using the traditional architecture.

5.3.4.3   High-Resolution SAR-ADC Redundancy Technology

Another reason for using redundancy is not the number of conversion clock cycles, but the number of reference voltages. Thus, instead of using a voltage reference with index 2, we can use a base of 1.85, for example, and then scale all references and capacitors proportionally.

Generally, in a binary ADC, the comparator compares V_{in} with $V_{ref}/2$ and then makes a decision. When the input signal is very close to the reference voltage ($V_{ref}/2$), any offset or noise in the comparator may cause the comparator to judge the wrong direction. In order to prevent small errors from affecting the conversion results, the use of redundant converters and codes with a base less than 2 (non-binary) should be considered. In the redundant converter shown in Fig. 5.30, if $V_{in} > V_{ref}/2 - $ err (e.g., err $= 12.7\% \, V_{ref}$), it is judged as 1, and if $V_{in} < V_{ref}/2 + $ err, it is judged as 0. Every judgment including wrong voltage can avoid the above problems. However, due to redundancy, it requires additional clock cycles.

In the design example presented in Fig. 5.31, the 10-bit A/D converter [22] requires 10 clock cycles. Assuming that we want to compensate for an error of 6.4%, we must add a clock cycle. If 12.7% is to be compensated for, 12 clock cycles must be introduced. However, significant differences can be observed in the

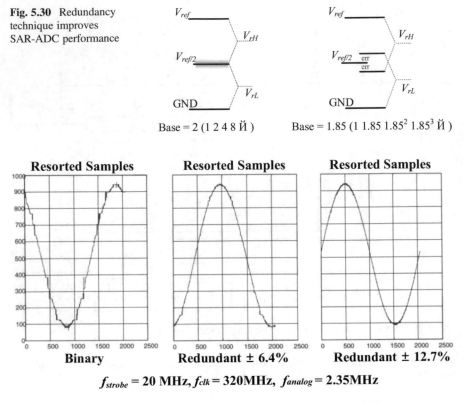

Fig. 5.30 Redundancy technique improves SAR-ADC performance

Base = 2 (1 2 4 8 Й)   Base = 1.85 (1 1.85 1.85^2 1.85^3 Й)

f_{strobe} = 20 MHz, f_{clk} = 320MHz, f_{analog} = 2.35MHz

Fig. 5.31 SAR ADC performance improvement with redundancy technique [22]

performance comparison. Compensating for 6.4% of errors is better than binary; it is better to compensate for 12.7% of errors.

5.3.4.4 High Sampling Frequency

The conventional implementation of the SA algorithm relies on clock synchronization to divide the time into a signal sampling phase and a conversion phase from MSB to LSB. The synchronization method requires the clock to run at least $(N + 1)$ $\cdot f_s$. For high-speed converters, the synthesis of such a high-frequency clock plus the clock distribution network may consume more power than the ADC itself. In terms of speed, since the decision time of the comparator largely depends on the input signal, when the input of the comparator is very small, the decision time of the comparator is relatively long. The clock speed is designed based on the worst-case clock cycle, including the maximum DAC setting time and the comparator resolution time. Asynchronous processing using internal comparison eliminates the need for such clocks and significantly improves power efficiency. Since the concept of the clock A/D converter is similar to a domino, after a decision is made at each level, the

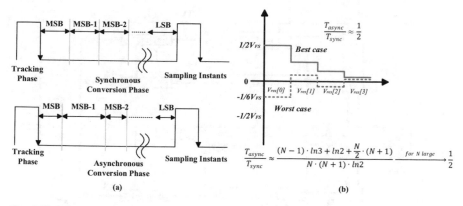

Fig. 5.32 High sampling frequency SAR-ADC asynchronous controller [1]

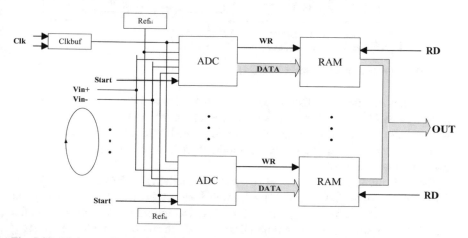

Fig. 5.33 High sampling frequency SAR-ADC interleaved architecture [23]

operation of the next level immediately starts, as presented in Fig. 5.32(a) [1]. Therefore, no high-speed clock is required.

As presented in Fig. 5.32(b), as N increases, the overall T_{async}/T_{sync} ratio reaches 1/2. The reduction in the maximum resolution time between synchronous and asynchronous cases is twofold.

5.3.4.5   Parallelism Improves Throughput

To obtain high throughput with low power consumption, parallelism should be applied, the energy efficiency of the simple ADC structure should be taken advantage of, and eight successive approximation ADCs should be used in parallel.

Figure 5.33 presents a block diagram of the chip architecture: eight converters trigger work in a time-interleaved period [23]. Since a single converter requires two

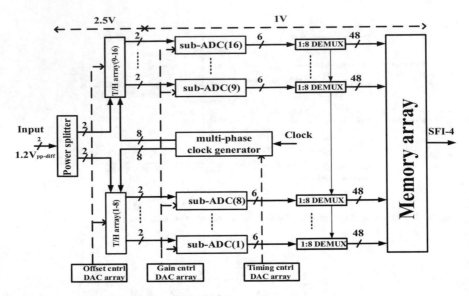

Fig. 5.34 A 24 GS/s 6b ADC architecture [24]

clock cycles for data acquisition and six clock cycles for conversion, the total sampling rate is equal to the clock rate. The on-chip RAM should be used as a buffer memory, and the converted data should be read. Multiple "slow" SAR-ADCs should be interleaved to obtain high throughput while maintaining low complexity and good power efficiency.

The ADC introduced in Fig. 5.34 [24] uses 16 interleaved 1.5 GS/s 6b sub-ADCs. Using the previous T/H circuit, the sampled signal is divided into two eight sub-ADC arrays at a sampling rate of 24 GS/s. The output of the sub-ADC is fed to the DEMUX array to generate time-aligned data to the memory at a write rate of f_s/64.

Please note that there are three digitally controlled calibration circuits at the bottom of Fig. 5.34 to correct the offset, gain mismatch, and timing deviation between the 16 channels. If each channel has a different offset, causing the output to contain tones introduced by a periodic error sequence, the sequence will produce spurs in the output spectrum. If the gain of the channel is slightly different, it will cause amplitude modulation or phase shift. If the phase of each channel clock does not match, it is the result of phase modulation (similar to aperture uncertainty).

5.4 ADC Simulation and Testing

Usually, after we design an amplifier circuit according to the specification, we want to know whether the behavior of the amplifier is consistent with our design. At this time, we will pour a sine wave from the input to see if its output meets our expectations. The periodic output of the amplifier should be checked, and Fourier analysis of the output should be conducted. Fourier coefficients provide us with information about distortion bandwidth, etc. We hope to do similar things with ADCs. However, the output of the ADC is only the result of quantizing the discrete-time series. Calculating the distortion bandwidth and other information from these discrete-time series is a bit complicated and requires digital signal processing skills to be completed.

5.4.1 Basics of Sampling, Quantization, and FFT

The period of the discrete-time signal is P, if

$$s[n] = s[n + P] \tag{5.40}$$

A periodic discrete-time series with P samples can be extended to a discrete Fourier series. However, unlike the continuous-time case, the discrete-time Fourier series consists of only p sine waves. It can be extended to discrete Fourier series:

$$s[n] = \sum_{k=0}^{k=P-1} A_k \cos \left[2\pi \frac{k}{P} n + \phi_k \right] \tag{5.41}$$

The discrete-time periodic signal with period P has only P coefficients in its Fourier expansion. The fundamental frequency of the Fourier series expansion is $2\pi/P$ radians per sample. The Fourier transform is just a method to quickly calculate the Fourier coefficients A_k, ϕ_k.

The period of P is usually called the FFT record length. Since sampling is generated by acquiring a continuous-time signal and sampling it with f_s, the fundamental frequency of the sequence is $2\pi \times P$ radians for each sample. When converted to Hz, it will become f_s/P Hz, and it is usually called the bin width of FFT.

Now, let us see how the input frequency generates the periodic output of the ADC. If the input frequency is represented by f_{in}, the sampling sequence is $2\pi f_{in}/f_s \cdot n$.

$$A \cdot \cos \left[2\pi \frac{f_{in}}{f_s} \cdot n \right] = A \cdot \cos \left[2\pi \frac{f_{in}}{f_s} (n + P) \right]$$

$$= A \cdot \cos \left[2\pi \frac{f_{in}}{f_s} n + 2\pi \frac{f_{in}}{f_s} P \right] \tag{5.42}$$

This sequence is periodic only when $2\pi f_{in}/f_s \cdot P$ is an integer multiple of 2π.

$$2\pi \frac{f_{\text{in}}}{f_{\text{s}}} P = m \cdot 2\pi \tag{5.43}$$

$$\frac{f_{\text{in}}}{f_{\text{s}}} = \frac{m}{P} \tag{5.44}$$

On the other hand, if $f_{\text{in}}/f_{\text{s}}$ is a rational number, the discrete sequence generated by sampling the continuous sine wave is periodic. This is an important deviation from continuous time in which any sine wave is periodic.

5.4.2   Selection of Input Frequency f_{in}

Therefore, another way of writing P/f_{s} is m/f_{in}, which means that in a record length, we must have an integer number of input cycles. Now that we know that f_{in} and f_{s} are rational numbers, how to choose the rational numbers is critical. Figure 5.35 illustrates the reasons for the choice.

Because any integers and fractions are collectively called rational numbers, if we simply choose $f_{\text{in}}/f_{\text{s}}$ as 1/4, we will obtain a pure sine wave with a sampling rate of 1/4 and pass a number with only three levels (1, 0, and -1). Now, if we take a pure sine wave and pass through the quantizer since the quantizer has only three levels, we expect to see a lot of distortions at the output. Let us see what happens with this specially selected input frequency. The input is $\cos[2\pi f_{\text{in}} t]$, when sampling with f_{s}:

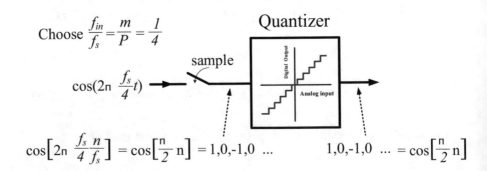

Fig. 5.35 Choice of f_{in}

$$\cos\left[2\pi f_{in}t\right] \xrightarrow[\text{sampling}]{f_s} \cos\left[2\pi\frac{f_s}{4}t\right] = \cos\left[2\pi\frac{f_s}{4}\frac{n}{f_s}\right] = \cos\left[\frac{\pi}{2}n\right]$$

$$\xrightarrow{\text{quanti.}} [1,0,-1,0\cdots1,0,-1,0\cdots] = \cos\left[\frac{\pi}{2}n\right] \qquad (5.45)$$

In this case, even though my quantizer is really bad, for a pure sine wave, the quantization distortion is zero. Something strange must have happened here, for the following reasons.

Mathematically, this can be understood by swapping the order of quantization and sampling operations. Quantization is a nonlinear operation that causes harmonics of the input sound to appear on the output. Thus, there will be f_{in}, $2\,f_{in}$, $3\,f_{in}$, and a bunch of harmonics. Subsequent sampling operations will alias many of these harmonics into the frequency range of 0 to $f_s/2$. Therefore, if the input frequency is chosen unwisely, many harmonics will alias into the same frequency band, making it appear as if there is no distortion. For the specific case where f_{in} is $f_s/4$, it is easy to see that all odd harmonics of the input frequency will be mixed into $f_s/4$. The third harmonic is $3f_s/4$, and the alias is $f_s/4$, which is the same as the input signal. This is why zero distortion will be observed even if the quantizer performance is poor. Therefore, the focus of the story is to select the input frequency so that important harmonics do not alias in the same place. To sum up the above results, $f_{in}/f_s = m/P$ must be a rational number. In order to prevent obvious harmonic aliasing, m and P should be selected (the numerator and denominator are relatively prime numbers). To calculate FFT, it is usually convenient to choose the record length (i.e., P) as a power of 2 (or $P = 2^Q$). Therefore, we usually set f_{in} to $f_{in} = (m/2^Q)$ f_s. When these conditions are met, the input tone will be located on the bin.

For example, if we want to test an 8-bit ADC running at 500 MS/s, but we want to test at $f_s/4$, we should not choose an input frequency of exactly 125 MHz. This will make a very bad converter look like a perfect converter. So, for example, a good choice is to select a record length of 1024 samples and make the numerator 251 m, we will obtain an input frequency of $f_{in} = (251/1024)500$ MHz $= 122.55$ MHz. It is also important not to round this number. It is unclear how many people have the opportunity to look at the ADC data sheet. If we look at the output spectrum, we will find that the input is some strange frequency, such as 122.55 MHz. The reason for choosing it is to make m and P relatively prime.

Now, we look at the left-side example of Fig. 5.36. When P is 32 and m is 8, what happens on the unit circle is to get 1, 2, 3, 4 samples, and then the 5th, 6th, 7th, and 8th samples. The same pattern repeats and keeps repeating. We keep receiving these four codes because f_{in} is actually $f_s/4$. But, we really do not want to keep receiving the same codes. We want to actually evaluate the ADC based on different codes. For the example on the right side of Fig. 5.36, where P is 16, when we choose m equal to 1, then this is obviously a prime number. When we rotate around the unit circle, all 16 samples will be obtained. After reaching 16, it will then repeat from 1. But in fact,

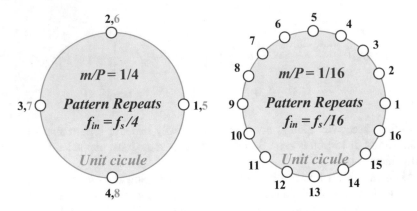

Fig. 5.36 Non-repeatable relative prime frequency

we will choose a more suitable prime number and distribute as many codes as possible on this unit circle, so that we can evaluate the actual behavior of the ADC.

5.4.3 FFT Leakage

The real meaning of calculating the FFT of a sequence of P points is as follows. We assume that this sequence is a period of a periodic discrete-time sequence with a period of P. In fact, we are calculating the coefficients of the discrete Fourier series of this periodic sequence, not the Fourier transform.

Let us see what happens when we input in the bin we discussed earlier. The input sine wave will have exactly an integer number of cycles within the record length. As you can see, if the input is not on the bin, there are no integer cycles, so there is a discontinuity in the periodic waveform, which is the period extension of these samples.

Whenever there is a jump in the time domain, it means that there is a lot of high-frequency energy, because there will be pulses. Furthermore, there is a lot of spectrum energy everywhere, which is called FFT leakage.

On the right side of Fig. 5.37, the blue spectrum is composed of the FFT of the sinusoid, which corresponds to the input tone in the upper left corner of the figure and is exactly on a bin, and only one of the FFT bins is nonzero. In the lower left corner, when the input is off a bin, we can see the energy leaks everywhere, forming the Eiffel Tower effect.

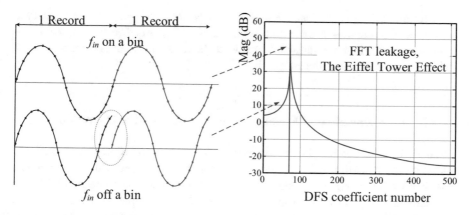

Fig. 5.37 FFT "leakage"

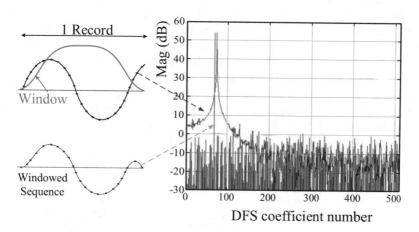

Fig. 5.38 Windowing effect

5.4.3.1   Causes of FFT Leakage

In the absence of leakage, the ADC output spectrum is shown in this example. Obviously, which is the input signal and which is the quantization noise. With regard to leakage, you can see that separating input noise and quantization noise is not easy.

The solution to FFT leakage is based on the fact that the problem lies at the boundary of the sampling record. This is because the discontinuity at the boundary of the record length is the cause of FFT leakage. This means that to solve this problem, we only need to focus on the middle of the record and reduce focus on the data at the boundary of the record. After the window function, we can practice the effect we want. Because the function is higher in the middle of the record, it gradually decreases to near zero at the end of the record. The input data sequence (where the input is not in the bin) is multiplied by the window function; the resulting sequence is presented in Fig. 5.38. When the sequence is regularly expanded, the discontinuities disappear. Therefore, when calculating the frequency spectrum of the sequence, the frequency spectrum components originally located in the high-frequency part will be significantly reduced. Such as the blue spectrum, that is, light leakage is greatly reduced.

5.4.3.2   Time-Domain and Frequency-Domain Windowing Operation

The measures that can be taken to mitigate the leakage should be explained. Refer to the leftmost illustration in Fig. 5.39 to consider the sine curve with frequency ω_0. This is an infinite sequence with Fourier transform, which is composed of two δ functions of $\pm\omega_0$. By looking at a finite number of P samples of an infinite sine wave, it is effective to multiply the sine wave by a rectangular window of width P in the time domain. Multiplication in the time domain is equivalent to convolution in the frequency domain. The Fourier transform of this P-length sequence is the convolution of the δ function and the sinc function centered on ω_0, which has a zero value at multiples of $2\pi/P$. Convolving these two functions will result in a Fourier transform centered on ω_0 and centered on ω_0, and there is a zero at $2\pi/P$ of ω_0. Refer to the far right illustration in Fig. 5.39.

The discrete Fourier series is obtained by sampling the discrete Fourier transform. Therefore, when calculating the FFT of these P samples, we are actually sampling the sinc function centered on ω_0. If ω_0 is on the bin, then we will sample the peak and zero points of the sinc function. Therefore, there is only one nonzero bin in the FFT. If the input is not just above the bin, it is called FFT leakage. When we sample the

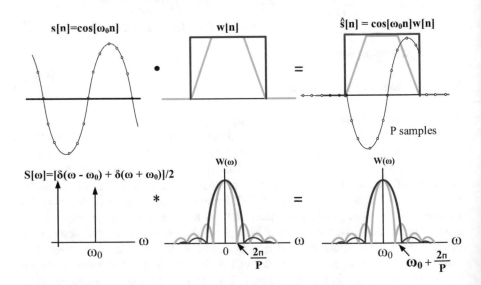

Fig. 5.39 Windowing operation in the time domain and frequency domain

Fourier transform, it is as if the nonzero shape |sinc| is depicted. Functionally, it looks like the Eiffel Tower.

As we saw before, the solution to this problem is to open a window. To make the intersection of the period extension zero or continuous, the weight of recording edge samples must be much smaller than the weight of recording intermediate samples. In addition, its edge in the time domain is slower than the rectangular window, which means that it is equivalent to a wider main lobe and lower side lobes in the frequency domain, indicating that it has lower high-frequency components.

5.4.3.3 Frequently Used Windows

Considering that the general idea of having windows is to weight the samples in the middle rather than the samples at the edges, Fig. 5.40 presents some commonly used windows [25]. A rectangular window is a window where you do nothing, and you will see that the high-frequency response is not very low. The Hann window, also known as raised cosine window, has a better high-frequency response, while the Blackman–Harris window has higher-frequency attenuation. These windows allow you to choose the trade-off between the main lobe width and side lobe suppression.

5.4.4 Testing the Nyquist ADC

If we plot the relationship between the average output code of a typical ADC and the DC input level, we see the true transmission characteristics of the ADC. Figure 5.41 presents the comparison between the actual ADC transfer curve and the idealized

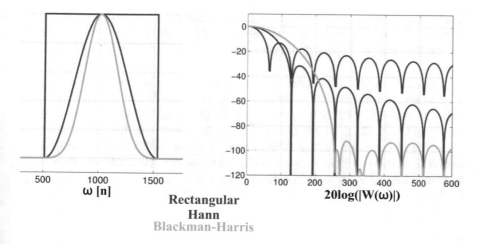

Rectangular
Hann
Blackman-Harris

Fig. 5.40 Common windows

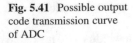
Fig. 5.41 Possible output
code transmission curve
of ADC

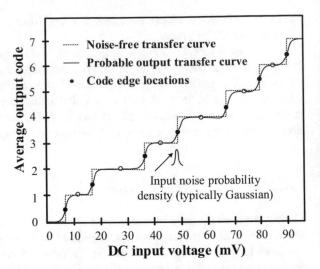

noise-free transfer curve [26]. The transition center (i.e., the decision-making layer) from one code to the next is usually called the coded edge. The wider the distribution of the Gaussian input noise, the smoother the transition from one code to the next. In fact, the true ADC transmission characteristic is equal to the convolution of the Gaussian noise probability density function and the noise-free transmission curve.

Once the coded edge is found, we can perform all the same tests on the ADC as those applied to the DAC. Generally, the code edge information is used for INL, DNL, DC gain, DC offset, and other tests.

5.4.4.1 ADC Code Edge Measurement

These two measurement methods are called center code test and edge code test. Figure 5.42 presents the difference between edge code testing and central code testing. The center of the code is defined as the midpoint between the edges of the code.

Figure 5.42 presents the central code test problem. Note that the center of the code is very close to a straight line, whereas the edges of the code exhibit much less linear behavior. Compared with edge code testing, the averaging process in the code center definition can produce artificially low DNL results. Therefore, this technique should be avoided. The edge coding method is a more discriminative test, so it is the preferred method to convert the transfer curve of the ADC into the one-to-one mapping required for INL and DNL measurements.

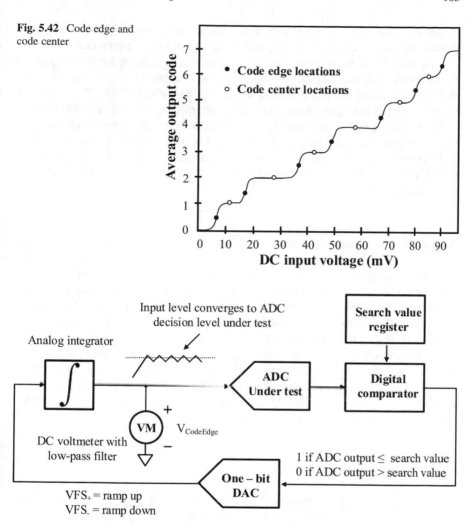

Fig. 5.42 Code edge and code center

Fig. 5.43 ADC servo test setup

5.4.4.2 Servo Method

The method of measuring code edges in production is to use servo circuits. Figure 5.43 presents a simplified block diagram of the ADC servo measurement setup. The ADC output code is compared with the value programmed into the search value register. If the ADC output is greater than or equal to the expected value, the integrator will ramp downward. If it is less than the expected value, the integrator will ramp upward. Finally, the integrator finds the desired code edge and fluctuates back and forth within its transition level. The average voltage $V_{CodeEdge}$ at the ADC input represents the falling edge of the code under test. This voltage can be easily measured using a DC voltmeter with an input that underwent LPF. The servo search process should be repeated for each code edge in the ADC transmission curve.

Although the servo method is easy to implement, it is also quite slow compared with the more common production test technique histogram method. The histogram test requires an input signal with a known voltage distribution. There are two commonly used histogram methods, namely, linear ramp method and sine method.

When testing ADCs, linearity is the most important performance indicator. The two most important errors are DNL and INL. We refer to code size errors as DNL and linear errors as INL. Both errors have been defined earlier in this chapter, and the definition can be found in JEDEC standards JESD 99-1 [27] and IEEE 1241-2000 [28] standards.

5.4.4.3   Linear Ramp Histogram Method

The ramp histogram is the simplest and most direct ADC linearity test method, as presented in Fig. 5.44. The stimulus signal should be a very linear ramp waveform. Its swing must be slightly larger than the ADC input range; otherwise, the linearity cannot be tested correctly. This is a very important point. The rising or falling speed of the ramp is set slow enough that each ADC code is "hit" multiple times. Since the histogram method is a statistical method, each code should appear multiple times, for example, at least 10 counts. Figure 5.44 presents the image with ramp codes. Because the input ramp is overloaded to the input range of the DUT, code 0 and full-scale code $(2^n - 1)$ appear much more frequently than the remaining codes. Here, n represents the number of digits.

5.4.4.4   Linear Ramp Equation

The linear calculation is as follows. Ignore the code 0 and code $(2^n - 1)$ counts, and add all the counts from code 1 to $(2^n - 2)$. The average height (H_m) of the histogram from 1 to $(2^n - 2)$ is calculated as follows:

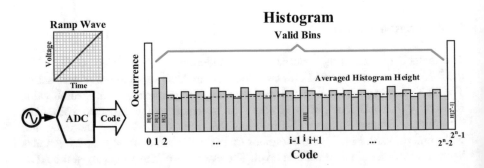

Fig. 5.44 Linear ramp histogram

$$H_{\mathrm{m}} = \frac{\sum_{i=1}^{2^n-2} H(i)}{2^n - 2} \tag{5.46}$$

Dividing H(i) by H_{m}, we obtain the width of each codeword in LSB as follows:

$$\text{Code Width}(i) = \frac{H(i)}{H_{\mathrm{m}}}, i = 1, 2, \ldots 2^n - 2 \tag{5.47}$$

It is necessary to exclude the highest and lowest code numbers because these two codes do not have a defined code width. In order to calculate the absolute or best-fit INL and DNL curves, we must determine the absolute voltage for each decision level. The differential linearity error DNL(i) is described as follows:

$$\text{DNL}(i) = \frac{H(i) - H_{\mathrm{m}}}{H_{\mathrm{m}}} = \frac{H(i)}{H_{\mathrm{m}}} - 1 \; [\text{LSB}] \tag{5.48}$$

where $i = 1,2,3,\ldots,2^n - 2$, and $\text{DNL}(0) = \text{DNL}(2^n - 1) = 0$.

The integral linearity error INL(i) is obtained as the accumulation of DNL(i), which is described as follows:

$$\text{INL}(i) = \sum_{k=1}^{i} \text{DNL}(i) \; [\text{LSB}] \tag{5.49}$$

Among them, it is known that $\text{INL}(0) = \text{INL}(2^n - 1) = 0$.

The possible results after measuring the slope histogram are presented in Fig. 5.45. The number of occurrences of each code is proportional to the width of the code. In other words, wide codes are hit more frequently than narrow codes.

5.4.4.5 Sine Histogram Method

The slope method is very simple in terms of linear calculation, because the slope histogram presents a flat linear profile. However, it is not easy to actually generate a good linear ramp waveform. A precision active integrator circuit with low loss and low dielectric absorption capacitors is required to generate a good slope. In traditional IC testing, high-precision DACs are generally used instead of integrators to generate pseudo-precision ramp waveforms. On the other hand, in the sine wave histogram method, a very low distortion sine wave is required. It is easier to generate a pure sine wave than to generate a perfect linear slope, because a suitable LPF can easily remove multiple harmonic distortions. Therefore, early testers usually rely on the sine histogram test of the high-resolution ADC. The second more common reason for using the sinusoidal histogram method is that it can better characterize the dynamic performance of the ADC. For this, we can use a high-frequency

Fig. 5.45 ADC sample
from the linear ramp
histogram test

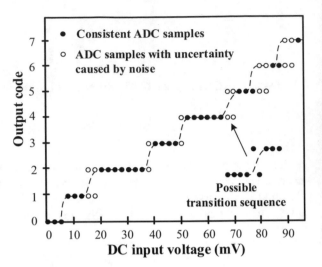

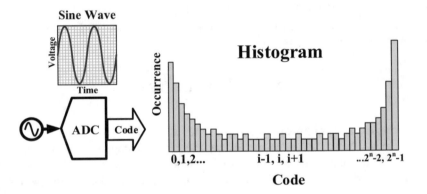

Fig. 5.46 Sine histogram

sinusoidal input signal. Our goal is to make the ADC respond to a rapidly changing
sinusoidal input rather than a slowly changing ramp voltage. However, since the
ADC generates a non-flat histogram distribution for the sine wave, as presented in
Fig. 5.46, the post-processing of the sine histogram for linear calculation becomes
much more complicated than the case of the slope histogram.

To clearly understand the details of this test setup, consider the illustration in
Fig. 5.46. The diagram consists of three parts. At the center is the transfer characteristic of the 4-bit ADC, that is, the 16 output code levels, expressed as a function of
the ADC input voltage. Below the ADC input voltage axis is a rotating graph of
ADC input voltage changes over time. The input signal here is a sine wave with peak
amplitude with DC offset and offset. These are based on the intermediate level of the
ADC input and are defined by the distance between the upper and lower decision
levels (V_{UE} and V_{LE}):

$$V_{\text{MID}} = V_{\text{LE}} + \rho, \text{and } \rho = \frac{V_{\text{UE}} - V_{\text{LE}}}{2} \tag{5.50}$$

According to the expression of the LSB step size V_{LSB}:

$$V_{\text{LSB}} = \frac{V_{\text{UE}} - V_{\text{LE}}}{2^D - 2} = \Delta \tag{5.51}$$

Therefore (5.51), we realized that we could also write the following:

$$\rho = \left(2^{D-1} - 1\right)\Delta \tag{5.52}$$

On the left side of the ADC transmission curve, we have drawn a histogram of the ADC code level (although rotated by 90°). Here, we see that the histogram has a shape like a "bathtub." If we try to use this histogram result in the same way as the linear slope histogram result, the upper and lower codes will be much wider than the middle code. Obviously, we need to normalize the histogram to eliminate the influence of the uneven voltage distribution of the sinusoidal waveform.

The normalization process is a bit complicated because we do not really know that the gain and offset of the ADC are a priori. In addition, we may not know the exact offset and amplitude of the sinusoidal input waveform. Fortunately, we have a piece of information that can tell us the level and offset of the signal seen by the ADC.

The number of hits in the upper and lower codes in the histogram can be used to calculate the offset and amplitude of the input signal. For example, from Fig. 5.47, we can see that the hit rate of the lower code is higher than that of the higher code. Since the sinusoid has a negative offset, it will hit the lower codes more frequently. The mismatch between these two numbers tells us the offset, whereas the number of total hits tells us the amplitude. Consider the pdf for the input voltage seen by the ADC

$$f(v) = \begin{cases} \dfrac{1}{\pi\sqrt{\text{peak}^2 - (v - \text{offset})^2}}, & -\text{peak} + \text{offset} \leq v \leq \text{peak} + \text{offset} \\ 0, \text{otherwise} \end{cases} \tag{5.53}$$

The probability that the input signal is less than the lowest code decision level now defined by $-\Delta$ (i.e., relative to the ADC intermediate level) is given by

$$P(V_{\text{IN,ADC}} < -\rho) = \int_{-\text{peak}+\text{offset}}^{-\rho} f(v)dv = \frac{1}{\pi}\left[\sin^{-1}\left(\frac{-\rho - \text{offset}}{\text{peak}} + \frac{\pi}{2}\right)\right] \tag{5.54}$$

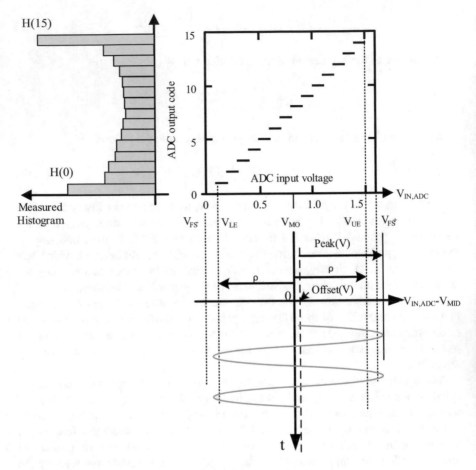

Fig. 5.47 Histogram of the sine of an ideal ADC

The probability that the input signal is greater than the highest coding decision level + Δ is calculated in a similar manner:

$$P(\rho < V_{IN,ADC}) = \int_{\rho}^{peak+offset} f(v)dv = \frac{1}{\pi}\left[\frac{\pi}{2} - \sin^{-1}\left(\frac{\rho - offset}{peak}\right)\right] \quad (5.55)$$

If N samples are collected from the ADC output (including the end code count), the expected code hits for code 0 and code 2^{D-1} are determined by the following equation:

$$H(0) = N \times P(V_{IN,ADC} < -\rho) = \frac{N}{\pi}\left[\sin^{-1}\left(\frac{-\rho - offset}{peak} + \frac{\pi}{2}\right)\right] \quad (5.56)$$

$$H\left(2^{D-1}\right) = N \times P(\rho < V_{\mathrm{IN,ADC}}) = \frac{N}{\pi}\left[\frac{\pi}{2} - \sin^{-1}\left(\frac{\rho - \text{offset}}{\text{peak}}\right)\right] \quad (5.57)$$

Here, we see two equations and two unknowns, which leads to the following solutions:

$$\text{offset} = \left(\frac{C_2 - C_1}{C_2 + C_1}\right)\rho = \left(\frac{C_2 - C_1}{C_2 + C_1}\right)\left(2^{D-1} - 1\right)V_{LSB} \quad (5.58)$$

and

$$\text{peak} = \left(\frac{2}{C_2 + C_1}\right)\rho = \frac{2}{C_2 + C_1}\left(2^{D-1} - 1\right)V_{LSB} \quad (5.59)$$

where

$$C_1 = \cos\left(\pi\frac{H\left(2^{D-1}\right)}{N}\right) \text{ and } C_2 = \cos\left(\pi\frac{H(0)}{N}\right) \quad (5.60)$$

It is very difficult to manually calculate INL/DNL. We generally use Matlab to help us calculate the measurement results into INL/DNL and plots. The Matlab code of the sine histogram method to calculate INL/DNL is shown in the appendix.

We should note that N should be large enough so that each ADC code is hit at least 16 times. The usual rule of thumb is to collect at least 32 samples for each code in the ADC transfer curve. For example, an 8-bit converter requires $2^8 \times 32 = 8192$ samples. Of course, due to the curved nature of the sinusoidal input, some codes will have more than 32 hits, whereas some codes will have less than 32 hits. Once we know the peak and offset values, we can calculate the ideal sine wave distribution.

Figure 5.48 presents an illustration of the histogram technique and sample size. In the upper-left corner of the picture is a 10-bit ADC; because this ADC has a

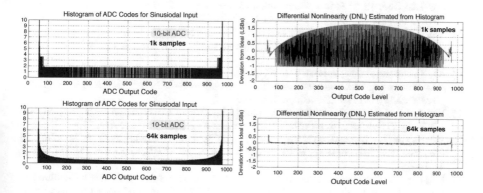

Fig. 5.48 Histogram technique and DNL vs. sample size [29]

thousand codes, we only execute it a thousand times. Therefore, on average, there is only one hit per bin, which is obviously not enough. Some of them have more than one hit rate, whereas others do not. It may look like a missing code, but it is not actually a missing code. Figure 5.48 presents the 64 K sample at the bottom left, and Fig. 5.48 shows the DNL for the same 1 k and 64 k samples.

References

1. Chen, S. W. M., & Brodersen, R. W. (2006). A 6-bit 600-MS/s 5.3-mW asynchronous ADC in 0.13-μm CMOS. *IEEE Journal of Solid-State Circuits, 41*(12), 2669–2680.
2. Peluso, V., et al. (1998). A 900-mV low-power Delta Sigma A/D converter with 77-dB dynamic range. *IEEE Journal of Solid-State Circuits, 33*(12), 1887–1897.
3. Tai, H.-Y., et al. (2014). A 0.85fJ/conversion-Step 10b 200kS/s Subranging SAR ADC in 40nm CMOS. In *ISSCC, 2014*, pp. 196–197.
4. Giannini, V., et. al. (2008). An 820uW 9b 40MS/s noise-tolerant dynamic-SAR ADC in 90nm digital CMOS. In *ISSCC, 2008*, pp. 238–239.
5. Kester, W. (2005). Which ADC architecture is right for your application. *EDA Tech Forum., 2* (4), 22–25.
6. ITRS road map. Retrieved from http://www.itrs.net/
7. Boyacigiller, Z., Weir, B., & Bradshaw, P. (1981, February). An error-correcting 14b/20 μs CMOS A/D converter. In *IEEE International Solid-State Circuits Conference. Digest of Technical Papers*, pp. 62–63.
8. Lu, T. C., Van, L. D., Lin, C. S., & Huang, C. M. (2011, September). A 0.5 V 1KS/s 2.5 nW 8.52-ENOB 6.8 fJ/conversion-step SAR ADC for biomedical applications. In *IEEE Custom Integrated Circuits Conference (CICC)*, pp. 1–4.
9. Mehr, I., & Dalton, D. (1999). A 500-MSample/s, 6-bit Nyquist-rate ADC for disk-drive read-channel applications. *IEEE Journal of Solid-State Circuits, 34*(7), 912–920.
10. Wong, Y. L., Cohen, M. H., & Abshire, P. A. (2005). A floating-gate comparator with automatic offset adaptation for 10-bit data conversion. *IEEE Transactions on Circuits and Systems I: Regular Papers, 52*(7), 1316–1326.
11. Abougindia, I. T., Cevik, I., Zghoul, F. N., & Ay, S. U. (2015). A precision comparator design with a new foreground offset calibration technique. *Analog Integrated Circuits and Signal Processing, 83*(2), 243–255.
12. Van der Plas, G., Decoutere, S., & Donnay, S. (2006, February). A 0.16 pJ/conversion-step 2.5 mW 1.25 GS/s 4b ADC in a 90nm digital CMOS process. In *IEEE International Solid State Circuits Conference-Digest of Technical Papers*, p. 2310.
13. Shu, Y. S. (2012, June). A 6b 3GS/s 11mW fully dynamic flash ADC in 40nm CMOS with reduced number of comparators. In *2012 Symposium on VLSI Circuits (VLSIC)* (pp. 26-27). IEEE.
14. Xu, Y., Belostotski, L., & Haslett, J. W. (2011, June). Offset-corrected 5GHz CMOS dynamic comparator using bulk voltage trimming: Design and analysis. In *IEEE 9th International New Circuits and systems conference*, pp. 277–280.
15. Yao, J., Liu, J., & Lee, H. (2010). Bulk voltage trimming offset calibration for high-speed flash ADCs. *IEEE Transactions on Circuits and Systems II: Express Briefs, 57*(2), 110–114.
16. Verbruggen, B., Wambacq, P., Kuijk, M., & Van der Plas, G. (2008, June). A 7.6 mW 1.75 GS/s 5 bit flash A/D converter in 90-nm digital CMOS. In *IEEE Symposium on VLSI Circuits*, pp. 14–15.
17. Miyahara, M., Asada, Y., Paik, D., & Matsuzawa, A. (2008). A low-noise self-calibrating dynamic comparator for high-speed ADCs. In *2008 IEEE Asian Solid-State Circuits Conference*, pp. 269–272.

18. Ginsburg, B. P., & Chandrakasan, A. P (2005). An energy-efficient charge recycling approach for a SAR converter with capacitive DAC. In *IEEE international symposium on circuits and systems*. pp. 184–187.
19. Liu, C.-C., et al. (2010). A 10-bit 50-MS/s SAR ADC with a monotonic capacitor switching procedure. *IEEE Journal of Solid-State Circuits, 45*(4), 731–740.
20. Zhu, Y., et al. (2010). A 10-bit 100-MS/s reference-free SAR ADC in 90-nm CMOS. *IEEE Journal of Solid-State Circuits, 45*(6), 1111–1121.
21. Tong, X., & Zhang, Y. (2015). 98.8% switching energy reduction in SAR ADC for bioelectronics application. *Electronics Letters, 51*(14), 1052–1054.
22. Kuttner, F. (2002). A 1.2 V 10b 20MSample/s non-binary successive approximation ADC in 0.13μm CMOS. In *IEEE International Solid-State Circuits Conference. Digest of Technical Papers (Cat. No. 02CH37315), ISSCC*, pp. 176–177.
23. Draxelmayr, D. (2004). A 6b 600MHz 10mW ADC array in digital 90nm CMOS. In *2004 IEEE International Solid-State Circuits Conference (IEEE Cat. No. 04CH37519)*, pp. 264–265.
24. Schvan, P., et al. (2008). A 24Gs/s 6b adc in 90nm cmos. In *2008 IEEE International Solid-State Circuits Conference-Digest of Technical Papers*, pp. 544–634.
25. Burns, M., & Roberts, G. W. (2001). *An introduction to mixed-signal IC test and measurement. Vol. 2001*. Oxford university press.
26. Davidson, S. (2013). *An introduction to mixed-signal IC test & measurement*. IEEE Design & Test.
27. (1989). *Terms, definitions, and letter symbols for analog-to-digital and digital-to-analog converters*. JEDEC Standard no.99 Addendum no.1 (JESD99-1), Electronic Industries Association.
28. (2000). *IEEE Standard for Terminology and Test Methods for Analog-to-Digital Converters*. IEEE Std 1241-2000.
29. Buchwald, A. (2010). *Specifying and testing ADCs*. ISSCC Tutorial.

Chapter 6
A 0.3 V 10b 3 MS/s SAR ADC with Comparator Calibration and Kickback Noise Reduction for Biomedical Applications

This chapter will introduce a 10-bit successive approximation analog-to-digital converter (ADC) that operates at an ultra-low voltage of 0.3 V and can be applied to biomedical implants. The study proposes several techniques to improve the ADC performance. A pipeline comparator was utilized to maintain the advantages of dynamic comparators and reduce the kickback noise. Weight biasing calibration was used to correct the offset voltage without degrading the operating speed of the comparator. The incorporation of a unity-gain buffer improved the bootstrap switch leakage problem during the hold period and reduced the effect of parasitic capacitances on the digital-to-analog converter. The chip was fabricated using 90-nm CMOS technology. The data measured at a supply voltage of 0.3 V and sampling rate of 3 MS/s for differential nonlinearity and integral nonlinearity were +0.83/−0.54 and +0.84/−0.89, respectively, and the signal-to-noise plus distortion ratio and effective number of bits were 56.42 dB and 9.08 b, respectively. The measured total power consumption was 6.6 μW at a figure of merit of 4.065 fJ/conv.-step.

6.1 Introduction

The progress of the semiconductor industry has promoted the development of wireless sensor networks. Wireless sensor networks are composed of a series of distributed micro-power sensor nodes that can sense, process, and relay data to a central base station. The wearable biopotential sensor node can be used to monitor vital signs, such as an electrocardiograph (ECG), electroencephalogram (EEG), respiratory frequency, and blood oxygen, all of which have different bandwidths and dynamic ranges. The power reduction characteristics of SAR ADCs are usually utilized. The purpose of this work is to design an analog-to-digital converter suitable for multi-channel neural signal recording circuits to collect data. The collected neural

© The Author(s), under exclusive license to Springer Nature Switzerland AG 2022
C.-C. Hung, S.-H. Wang, *Ultra-Low-Voltage Frequency Synthesizer and Successive-Approximation Analog-to-Digital Converter for Biomedical Applications*, Analog Circuits and Signal Processing, https://doi.org/10.1007/978-3-030-88845-9_6

signals can be transmitted to external receivers through the frequency synthesizer in the previous chapter to provide a basis for disease diagnosis or they can be used with neural stimulation circuits to give appropriate stimulation signals in real time. It consists of 64 independent readout channels and an analog multiplexer. Adding an ADC and digitizing the signal will improve the functionality of the system and simplify data transmission. Since the useful frequency range in neural applications is limited to approximately 20 kHz, a single converter with a sufficiently high conversion rate can be used for time interleaving to read and digitize data from all channels. The requirements of the designed circuit are 10-bit resolution and a typical conversion rate of 3 MS/s.

There are some innovations we applied to the 10-bit SAR ADC applicable for biomedical implants or smart dust sensors. A pipeline comparator was utilized to maintain the advantages of dynamic comparators and reduce the kickback noise. Weight biasing calibration was used to correct the offset voltage without degrading the operating speed of the comparator. The incorporation of a unity-gain buffer improved the bootstrap switch leakage problem during the hold period and reduced the effect of parasitic capacitances on the digital-to-analog converter.

6.2   Proposed Design Techniques

Regarding the choice of ADC technology at low operating voltages, operational amplifiers in pipelined ADCs or sigma-delta ADCs typically use one or two high-gain stages for accurate residue amplification or integration [1–3]. In deep submicron CMOS, the gain $g_m \cdot ro$ achievable per stage is limited because the short channel effect reduces the ro value of a single transistor and the reduction in the supply voltage V_{DD} further reduces the available transconductance. Moreover, successive approximation (SAR) ADCs more effectively handle multiplexed inputs than incremental Σ–Δ ADCs. SAR ADCs do not require much post-signal processing, especially when only low oversampling rates are allowed. In sensor applications, events occur sporadically, and nodes may acquire data only once before they must react. Thus, an SAR ADC with the ability to convert one sample at a time is suitable. Compared with operational amplifiers, the comparators used in SAR ADCs only require a small output swing to distinguish between decisions from noise. The supply voltage scaling in advanced CMOS processes has less effect on the comparator's judgment, but the sensing development time is long. To maintain a fast response time, the use of low-threshold components is necessary [4, 5]. In summary, the SAR ADC technology is selected because of its advantages, such as low power consumption, high resolution, small size, and high speed.

The main sources of energy consumption in SAR ADCs are SAR logic, comparators, and capacitive DACs [6]. Reducing the value of the unit capacitance Cu to the limits allowed by the technology and limiting the sampling noise kT/C reduces the energy consumption. Furthermore, the use of a separate capacitor array instead of a full binary-weighted array reduces the energy consumption by reducing the total

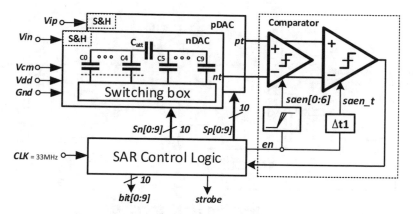

Fig. 6.1 Overall schematic diagram of the ADC architecture

capacitance of the array and improves the problematic large ratio between the
capacitors to achieve a good match. Further energy conservation can be achieved
using effective switching methods [7–11]. The method used in this chapter is the
V_{cm}-based switching approach, which reduces the capacitance of the MSB. Moreover, the switching sequence reduces the energy consumption by an additional value
of 1/3 [12] by using the set-and-down technique [7]. By changing the common-mode
voltage of a comparator during the comparison process, the comparator is caused to
generate an additional offset in the comparator, which affects the linearity of the
ADC. The V_{cm}-based switching method can also prevent such problems.

The block diagram of the SAR ADC proposed in this chapter is presented in
Fig. 6.1. The figure presents the architecture of a 10-bit split-SAR ADC with V_{cm}-
based switching. A differential DAC output is connected to the two-stage pipeline
comparator. The first stage is a dynamic comparator with controlled slew rate and
digital calibration techniques and the second stage is a high-speed dynamic comparator. The bootstrap-type sampling switch performance is enhanced by incorporating
a new feedback circuit to reduce the leakage problem. A shielding layer is used to
reduce the influence of the parasitic capacitance of the top plate on linearity. The
input clock of 33 MHz with one sampling and ten conversions achieves a sampling
rate of 3 MS/s.

6.2.1 Two-Stage Pipeline Comparator

To amplify a small input voltage to a full logic voltage, a comparator requires very
high gain. An example of using a static amplifier as a preamplifier to reduce the
offset voltage is presented as follows: two stages of static amplifiers were connected
in [13]. Each stage has a gain of less than 10 and is enabled only during amplification
to reduce the static power consumption. The bias condition of the static amplifier has

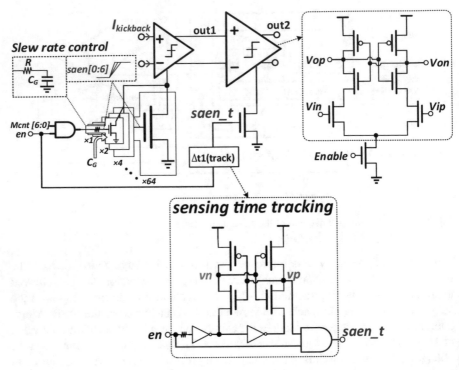

Fig. 6.2 Pipeline comparator with slew rate control and sensing time tracking

to be re-established and a transient noise is generated when the power is switched on. Compared with static amplifiers, the gain of dynamic comparators is $\exp(\Delta t/\tau_m)$ due to the positive feedback, where $\tau_m = C/g_m$ is the regeneration time constant, C is the load, and g_m is the transconductance. As the gain increases exponentially with time, achieving a gain of more than 10 and even 100 is easy while consuming low power to replace the static preamplifier. However, the issue with the dynamic comparator is that it presents the kickback noise. This is because when the regeneration phase begins, the switch is switched on and the two cross-coupled inverters implement a positive feedback, which causes the output voltage to enhance from the original reset phase. The small original output voltage tends to reach zero or V_{DD}. A large voltage change on the regenerative node is coupled to the input of the comparator through the parasitic capacitance of the transistors. As no zero output impedance of the capacitor array exists in the input of the comparator, the voltage on the capacitor array is disturbed, which may degrade the accuracy of the converter. The kickback noise is proportional to the regenerative speed which can be controlled by the slew rate of the gate voltage of the tail transistors in the dynamic comparator. Figure 6.2 presents the kickback noise reduction by adjusting the gate voltage slew rates of the tail transistors in the first stage of the comparator.

For the slew rate control, the first stage of the comparator uses the binary-weighted gate capacitance C_G of the binary-sized tail transistors and serially

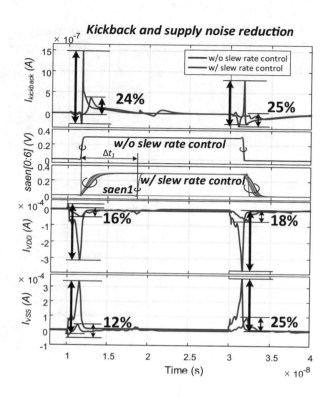

Fig. 6.3 Kickback and supply noise reduction by slew rate control

connected resistors (implemented by pass gates) with their gate terminals to form slow slew rate control which reduces about 75% kickback noise at the rising and falling edges of gate voltage in Fig. 6.3. Also because the conduction sequence is from small-sized to large-sized tail transistors, a smoother power/ground current sink is obtained and the ($L \cdot di/dt$) noise is decreased. Some previous designs enable the two stages of the comparator simultaneously for fast response time [14–18], but instantaneously large sink current generates large power/ground line noise through coupling (substrate coupling, power line noise coupling, or parasitic capacitor coupling....), interferes the small signal to be sensed, and may cause comparator misjudgment which is un-recoverable because of positive feedback mechanism of the comparator. Separating the sensing time (or signal development time) between two stages by $\Delta t1$ results in less power/ground noise to avoid from erroneous decisions. Because of the slew rate control, the peak power and ground spike current reduce up to 80% and 75% for sensing rising and falling edges.

To ensure the enough gain of the design, the first stage of the comparator starts to operate first, so *saen[0:6]* first goes high sequentially. Then, *vn* and *vp* in the sensing time tracking circuit fight with two inverters' outputs, start to transit their states based on the same mechanism of the cross-coupled latch in the comparator, and finish transitions after $\Delta t1$, resulting in *sane_t* becoming high, which starts the second-stage operation as shown in Fig. 6.4, where the input difference of the comparator is 150-μV [0.5 least significant bit (LSB)]. $\Delta t1$ in TT and SS corners

Fig. 6.4 Comparator outputs with sensing time tracking circuit for different corners

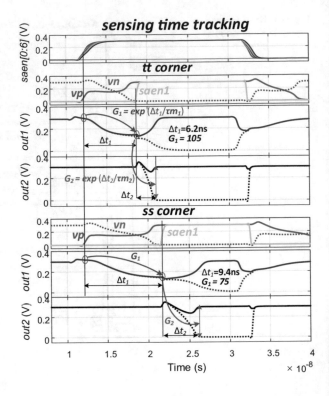

Fig. 6.5 Pipeline comparator sensing speed vs. supply voltage

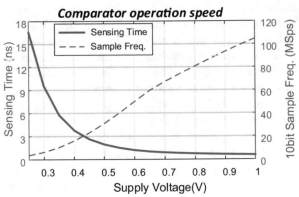

are 6.2 ns and 9.4 ns, respectively, and the gains are 105 and 75, respectively. After $\Delta t1$, then the second stage follows to operate.

Figure 6.5 estimates the operation speed of the 10-bit SAR ADC (11 cycles including one sampling period) by considering the operating speed of the comparator (including the required reset time); the ratio of the sensing time to the reset time of the comparator is 2:1. The pipeline comparator can operate at a typical corner condition with a sampling speed of 5 MS/s at 0.3 V and a sampling speed of

100 MS/s at 1.0 V. This enables the wide use of pipeline comparators in ADC applications with different sampling speeds.

6.2.2 Weight Biasing Calibration

Two types of comparator mismatches can occur: (1) static mismatch due to changes in the current factor β ($= \mu_n C_{ox} W/L$) and the threshold voltage V_T and (2) dynamic mismatch due to some internal node capacitor imbalance [19, 20] that results in the offset voltage of the comparator and cannot eliminate itself; mismatches must be corrected with calibration. The two types of comparator calibration are background and foreground calibrations. Background calibration is performed during normal ADC operation, but it requires at least one additional clock cycle for each ADC conversion. However, its implementation presents potential problems because it is required to reduce the voltage across the input terminals of the comparator from any possible residual value on the CDAC top plate to a 0 V within one clock cycle before the calibration. The load effect of the transistor source–drain junction capacitor that is in parallel with the CDAC top plate during normal operation presents a channel charge release caused by switching. Moreover, the leakage current between the source and drain affects its accuracy, and the leakage is even worse under a low supply voltage. In the foreground calibration, an offset correction was conducted during ADC startup. Hence, it is relatively easy to implement and is suitable for most ADC topologies. For biomedical implant systems or multifunction smart dust sensors, a sleep mode designed to save energy and a foreground calibration that performs an offset correction during startup are suitable.

The proposed concept of the weight biasing calibration resembles the "weight adjustment" mechanism on the tray balance, thus allowing the originally unbalanced balance to return to equilibrium, as illustrated in Fig. 6.6(a). Before the comparator begins the comparison, appropriate weights are assigned to compensate for the bias or offset of the comparator and achieve an equilibrium.

The first stage of the comparator is illustrated in Fig. 6.6(b). The drain–source current I_{DS} for N-type MOSFET is

$$I_{DS} = K \cdot (V_{GS} - V_T)^2; K = \frac{1}{2} \mu_n C_{ox} \frac{W}{L} \tag{6.1}$$

where V_{GS} is the gate-source voltage, V_T is the transistor threshold voltage, μ_n is surface mobility of electrons, C_{ox} is the capacitance per unit area of the gate electrode, W is channel width, and L is channel length. The offset voltage Vos of the comparator circuit presented in Fig. 6.6(b) is where the transistor current of M_1, I_{DS1}, is equal to the transistor current of M_2, I_{DS2}. When the two currents are equal, I_{DS1} and I_{DS2} become equal to half the tail current I_{tail}.

By assuming that I_{DS1} and I_{DS2} are equal, the offset voltage Vos for the comparator is

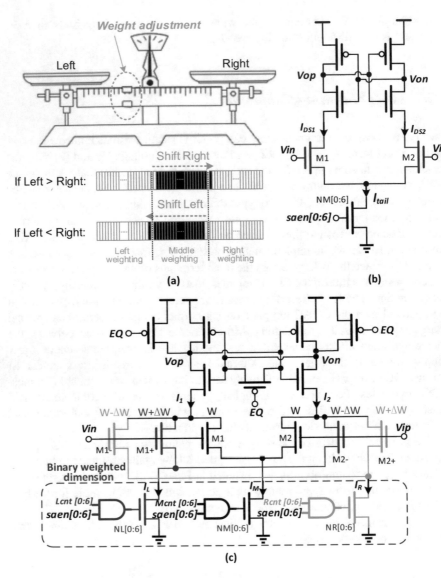

Fig. 6.6 (a) Tray balance with weight adjustment, (b) first stage of the comparator, and (c) first stage of the comparator with weight adjustment

$$\boldsymbol{Vos} = V_{GS1} - V_{GS2} \tag{6.2}$$

Solving Eq. (6.1) for V_{GS} and assuming that I_{DS1} and I_{DS2} are both equal to $I_{tail}/2$, Eq. (6.2) can be rewritten as

$$Vos = \sqrt{\frac{I_{\text{tail}}}{2K_1}} - \sqrt{\frac{I_{\text{tail}}}{2K_2}} \tag{6.3}$$

where K_1 and K_2 are the constants of M_1 and M_2, respectively. From Eq. (6.3), it is evident that if transistors M_1 and M_2 are equal (i.e., $K_1 = K_2$), the offset voltage becomes zero, and the offset increases with increase in the difference. According to Eq. (6.1), a preferred approach to make K_1 and K_2 different is by changing the channel width W. Equation (6.3) can be rewritten as follows:

$$Vos = \sqrt{\frac{I_{\text{tail}}}{2K_1}} \left(1 - \sqrt{\frac{K_1}{K_2}}\right) \tag{6.4}$$

If we have the value of Vos, K_1 and designed I_{tail}, then.

$$K_2 = \frac{K_1}{\left(1 - Vos\sqrt{\frac{2K_1}{I_{\text{tail}}}}\right)^2} \tag{6.5}$$

From Eq. (6.4), it is evident that for the given different ratios of K_1 and K_2, the offset voltage Vos of the comparator can be cancelled out.

Based on the variation in the process, the offset voltage may be positive or negative, that is, the bias to the left or right. Thus, we place two groups' input stages in parallel to the original input stage, which are right (green) and left (red) biases for the compensation of the original input stage, as illustrated in Fig. 6.6(c). The width of this compensation input stage (K_2) can be calculated by solving Eq. (6.5), while the maximum offset voltage value of the process was attained from Monte-Carlo simulation. The offset is controlled using the three sets of binary bits at the input stage. Specific current controllers (NL[0:6], NM[0:6], and NR[0:6]) are used to control the specific gravity of each group, where the value of the counter represents the magnitude of the bias current. To maintain the operating conditions of the original circuit, the total current flowing through the three sets of current controllers remains constant ($I_{\text{tail}} = I_L + I_M + I_R$). Only the weighting amounts are moved left or right. In the initial stage, $I_L = I_R = 0$ and $I_{\text{tail}} = I_M$. For example, to shift to the right, the amount of increase in I_R (NR + 1) is equal to the amount of decrease in I_M (NM − 1) to keep I_{tail} constant. Because the Monte-Carlo simulation indicates a maximum offset voltage of ±20 mV, the calibration procedure for the 0.3 V 10-bit ADC requires 20-mV/150-µV [0.5 least significant bit (LSB)] = 133 counts; a 7-bit (127) counter is used to simplify the design.

The calibration procedure is illustrated in Fig. 6.7. In the initial phase, all weights are at Mcnt = 127, that is, $I_M = I_{DS1} + I_{DS2} = I_{\text{tail}}$ and $I_L = I_R = 0$. Then, the weighting factor of Mcnt is shifted to the left or right based on the comparison. As it is only moving weights, the total bias current remains the same. The process is repeated 127 times to complete the calibration.

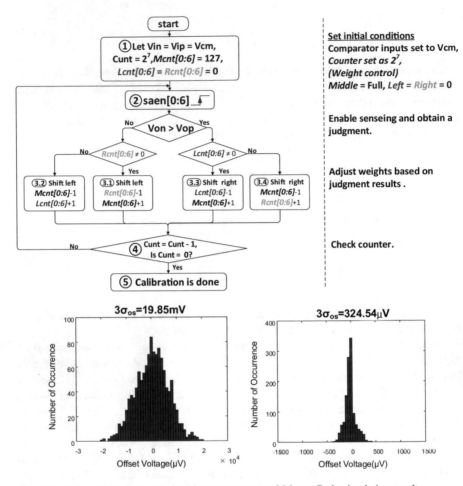

Fig. 6.7 Offset calibration procedure for comparator and Monte-Carlo simulation results

Monte-Carlo simulation was used to determine the variation in the process at the input stage. After completing the 127 cycles and adjusting the biasing weight based on the aforementioned calibration procedure, as presented by the waveforms in Fig. 6.8, the offset of the calibrated comparator was verified. Figure 6.7 also shows the results of 1200 Monte-Carlo samples of the post layout static offset voltage distribution before and after correction. The simulation results in Fig. 6.7 present that the three-sigma offsets before and after correction are 19.85 mV and 324.54 µV, respectively.

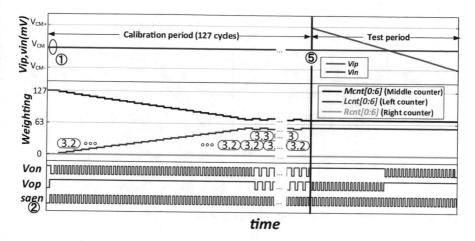

Fig. 6.8 Example of simulation waveforms of the calibration procedure

6.2.3 Bootstrap Switch Subthreshold Leakage Reduction

At low supply voltage, the challenge presented by the MOSFET transistor to achieve a good sampling switch is to increase the ratio of "on" conductance to "off" leakage current. The sampling switch must have an adequately high "on" conductance and/or linearity in the rail-to-rail range such that it does not introduce distortion. Moreover, the "off" current must not cause input-dependent ADC errors. The cause of leakage of the sampling switch is the subthreshold leakage, where the gate-to-source voltage of the MOS transistor is lower than the threshold voltage and the transistor is operating in the subthreshold region. The subthreshold leakage is a small amount of leakage current between the source and the drain.

The subthreshold leakage current I_{sub} was calculated using the following formula [21]:

$$I_{sub} = I_0 e^{\frac{V_{GS}-V_T}{nU_T}}\left(1 - e^{\frac{-V_{DS}}{U_T}}\right) \tag{6.6}$$

where $I_0 = \frac{W\mu_0 C_{ox}V_T^2 e^{1.8}}{L}$, $U_T = \frac{KT}{q}$ (U_T is the thermal voltage), V_T is the threshold voltage, and V_{DS} and V_{GS} are the drain-to-source and gate-to-source voltages, respectively. W and L are the effective transistor width and length, respectively. C_{ox} is the gate oxide capacitance, μ_0 is the carrier mobility, and n is the subthreshold swing coefficient. According to this formula, even when V_{GS} is set to 0 V, a current flows into the channel of the "OFF" MOSFET transistor because of drain and source terminals with V_{DS}. We can discern that greater V_{DS} corresponds to higher I_{sub} value by evaluating the aforementioned formula and focusing solely on the last term. I_{sub} can be minimized or even approximately made zero by setting $V_{DS} \approx 0$.

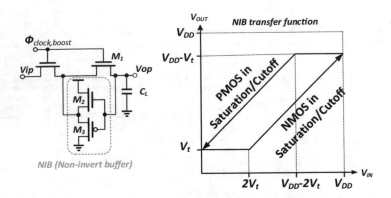

Fig. 6.9 Feedback leakage reduction and its transfer curve

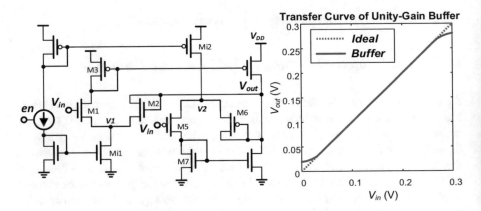

Fig. 6.10 Unity gain buffer and its transfer curve

The leakage feedback cancellation [22] is the most effective of the several methods, as illustrated in Fig. 6.9. However, three problems exist with low supply voltages: 1) the voltage swing at the output is limited, and the available swing range is $V_{DD} - 2 V_T$. 2) A voltage exists across the input and output of the buffer such that $V_{DS1} = |V_{GS2,3}| (\geq V_T)$. In this low-voltage application, $V_{DD} = 0.3$ V and $V_T = 0.25$ V, so $(1 - \exp(-V_{DS1}/U_T)) \approx 1$. Based on the transfer curve of the proposed unity-gain buffer illustrated in Fig. 6.10, $Vip = Vop$ can be used to make $V_{DS1} \approx 0$, which can suppress most of the leakage current, except a 20-mV region near the V_{DD}/Gnd voltage, where the term $(1 - \exp(-V_{DS1}/U_T)) \approx 0.55$ still can reduce the subthreshold leakage. 3) Because the non-invert buffer (NIB) is always maintained in an "on" state, it will result in a fighting problem. When sampling ($\overline{samp}$ goes low), the input signal will fight with the buffer output, thus distorting the input signal. The unity-gain buffer of this design can be controlled by the **en** signal. When the signal $\overline{samp}$ is connected to the **en** signal, it turns off the buffer to avoid

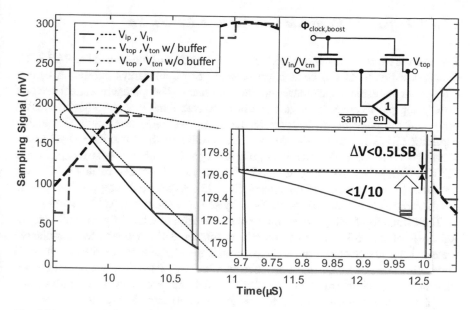

Fig. 6.11 Leakage reduction in the hold time period

interference with the input signal during sampling. Otherwise, it acts as a unity-gain buffer.

Figure 6.10 displays the proposed unity-gain buffer and its transfer curve. The unity-gain buffer is composed of the complementary differential input pairs and single-ended output and operates in the subthreshold region because the supply voltage is close to the threshold voltage. As the loading capacitor C_L is set to 20 fF and the slew rate (SR) request is 0.3 V/30 ns, $I_{bias} = SR \cdot C_L = 0.2$ μA and each $I_{DS} = I_{bias}/2 = 0.1$ μA. As $g_m/I_{DS} = 1/(nU_T)$ ($n \approx 1.5$, $U_T \approx 26$ mV), $g_m = 2.5$ μS. Therefore, a smaller size of input stage is acceptable with smaller input capacitance to alleviate the loading of the switch.

We replace the NIB in the sample and hold circuit of the bootstrap switch with the unity-gain buffer and check its validity by simulation, as illustrated in Fig. 6.11. According to the design, the leakage current of the sample and hold circuit during the hold period should limit the voltage deviation of the DAC capacitor to less than 0.5 LSB to satisfy ENOB. The simulation results show that the voltage drop of the unity-gain buffer during the hold period is approximately 80 μV. These results meet the 0.5 LSB (0.3-V / 2^{10} = 150-μV) design goal requirement for a 10-bit ADC. Compared with the circuit without unity-gain buffer, it reduces leakage by more than 80%.

6.2.4   Impact of Parasitic Capacitance Reduction

Because the capacitor array occupies the largest portion of the area in the SAR ADC circuit, the layout of the connection between capacitor arrays exhibits a certain amount of parasitic capacitance on the top plate. The parasitic capacitance of DAC's top plate causes only gain errors in conventional binary-weighted (CBW) array, but the parasitic capacitance in the binary-weighted array with attenuation capacitor (BWA) destroys the desired binary ratio of the capacitive DAC array. Thus, the conversion linearity is degraded. Regardless of the unitary capacitance (Cu) chosen, when the top plate parasitic increases above 6% of that of Cu, the converter loses one bit of effective resolution [23]. As the percentage of the top plate parasitic capacitance decreases, the linearity of the capacitive DAC increases.

The principle of using the equipotential operation to cancel the parasitic capacitance effect is as follows: If no potential difference exists between two conductive plates with parasitic capacitance, the parasitic capacitance between the two plates is "invisible" because no current flows between the two plates. The voltage of the top plate of the capacitor array is delivered to the shielding layer between the ground and top plate through unity-gain buffer, which is equivalent to isolate the parasitic capacitance of the top plate of the capacitor array from ground for improving the linearity of the DAC.

The simulated waveforms in Fig. 6.12 show the top plate signal of the LSB side of DAC and the shielding layer signal through unity-gain buffer. The parasitic capacitance of the shielding layer is approximately 19 fF (extracted by CAD tool). Figure 6.13 shows that the shielding layer is arranged under the metal–insulator–metal (MiM) capacitor, and the enable control signal *en* of the unity-gain buffer is connected to V_{DD}.

With or without enabling the equipotential circuit operation, the linearity simulation shows no significant difference in the maximum and minimum values of differential nonlinearity (DNL), as illustrated in the post simulation of Fig. 6.14. As the unity-gain buffer shows a voltage headroom (~ 6%) near V_{DD}/Gnd, the effect of parasitic capacitance cannot be compensated. Hence, the DNL maintains the original amplitude near the maximum and minimum digital codes while the DNL

Fig. 6.12 DAC top plate and unity-gain buffer output waveforms

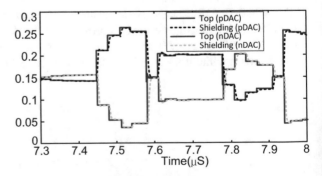

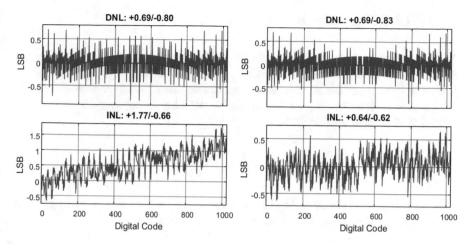

Fig. 6.13 Simulation of static performance with and without enabling the equipotential circuit

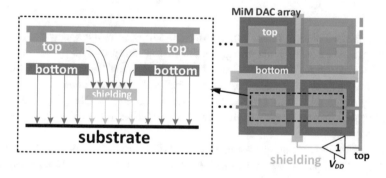

Fig. 6.14 Shielding layer under the MiM capacitor forms a equipotential layer to cancel the parasitic capacitance effect of the top plate of the capacitor array

in the other areas decreases. The accumulation of smaller DNL results in 1.17 LSB improvement in the entire integral nonlinearity (INL).

6.3 Circuit Implementation

The proposed circuit was fabricated by TSMC 90-nm CMOS process with MiM capacitors. The BWA structure can be applied to the 10-bit resolution targets by using MiM capacitors and DAC mismatch calibration [24, 25].

6.3.1   Unit Capacitor and DAC Topology

The effect of capacitance mismatch for the BWA structure can be quantified by the maximum standard deviations of the DNL ($\sigma_{DNL.max}$), which typically occur at the mid-code level. The combined relative standard deviation of the unit capacitance is analytically derived in [26], and $\sigma_{DNL.max} < 0.5$ LSB is required to prevent the influence of process mismatch. That is,

$$\sigma_{DNL,\,max} = 2^{\frac{3N}{4}} \cdot \sigma\left(\frac{\Delta C_u}{C_u}\right) < 0.5 \tag{6.7}$$

By considering the relationship between two equal unit capacitors and neglecting the size-independent term, the variation can be assessed as [27].

$$\sigma\left(\frac{\Delta C_u}{C_u}\right) = \sqrt{\frac{k_c^2 c_{spec}}{C_u}} \tag{6.8}$$

where kc is the Pelgrom mismatch coefficient and c_{spec} is the specific capacitance. The parameters $kc = 1.01\%$ and $c_{spec} = 2.0$ fF/um^2 are provided by the fabrication. For 10-bit resolution, $Cu > 27$ fF can be obtained using Eqs. (6.7) and (6.8). For considering the minimum available fractional value of the capacitor that can be accurately obtained, the unit capacitance was set to 44 fF. The bridge capacitance was calibrated using the method presented in [13]. Figure 6.15 illustrates the BWA DAC topology.

For considering the signal-to-noise ratio (SNR) limitation caused by the sampling noise kT/C, the input voltage was sampled using the sum of the sampling capacitors

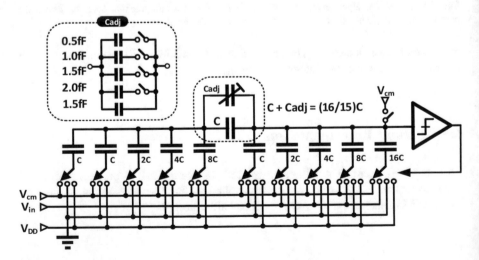

Fig. 6.15 BWA DAC topology

Cs. The on-resistance R_{on} of the switching transistor causes thermal noise. Simultaneously, the resistor forms a low pass filter with the sample capacitor. The noise can be calculated based on the formula presented in [28].

$$n = \sqrt{\frac{KT}{C_s}} \tag{6.9}$$

If fully differential sampling is considered, the calculated noise of the 1.4 pF sampling capacitor C_s is 54-µV rms or 76-µV rms peak. These values are lower than those for a 0.5 LSB at an input voltage of 0.3 V and 10-bit resolution; thus, the sampling noise can be neglected. As the capacitance for fractional calibration is very small, two layers of metal are used to implement the small capacitance. Figure 6.17 illustrates the DAC arrangement.

6.3.2 Signal and Noise Consideration

Although lowering the supply voltage is an effective way to reduce dynamic power consumption, the extremely low supply voltage decreases the signal amplitude, thus degrading the noise tolerance. A lower signal amplitude requires a lower noise floor. It is crucial to determine how to reduce the noise interference while maintaining signal stability without degrading the SNR.

The pipelined comparator technology presented in the previous section not only reduces the kickback noise generated by the comparator in the DAC, but also reduces the instantaneous bounce noise to the supply ring. Although [23] reveals that the addition of dummy capacitors causes parasitic effects in the capacitive DAC, the entire ENOB is reduced by approximately 0.2 bits. However, to reduce the mismatch and isolate the noise interference during the high voltage conversion of the bootstrap circuit, we added dummy capacitors and used those as V_{CM} decoupling capacitors to stabilize the reference voltage.

Compared with single-ended implementations, the differential implementations of the SAR ADC are relatively immune to noise and other factors that degrade ADC performance. The sampling noise can be suppressed using the large sampling capacitor. For bit cycling, the time required for reference setup or charge sharing of DAC is typically the bottleneck of the SAR ADC speed, especially when the chip provides a reference through the die bond wire. In this chapter, in addition to having an on-chip capacitor to stabilize each reference source, which suppresses the induced ringing effect for achieving faster reference stabilization, the switch in each bit has its own capacitor reservoir. As illustrated in Fig. 6.16, the bit local capacitor reservoir quickly supplies charge to the DAC and reduces the setup time for each bit cycle.

Fig. 6.16 Local capacitor reservoir quickly supplies charge to the DAC

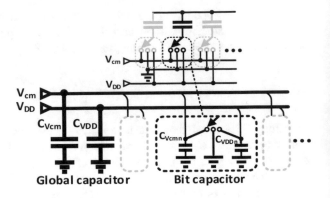

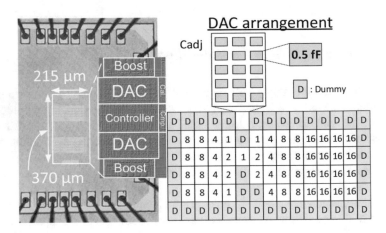

Fig. 6.17 Die photo and DAC arrangement

6.4   Measurement Results

The proposed SAR ADC was fabricated by TSMC 90-nm CMOS mixed-signal general-purpose 1P9M process technology with low-threshold devices. Figure 6.17 displays the photomicrograph of the chip. The chip area is approximately 0.0795 mm^2 (0.215 × 0.37 mm^2) by excluding the pad area. The analog input differential sine wave signals and digital clock signal were generated using the M9036A PXIe Embedded Controller (Keysight) and M9336A PXIe I/Q Arbitrary Waveform Generator (Keysight), respectively. A logic analyzer (16861A, Keysight) captured a total of 11 bits (10 data bits and a strobe signal) and a 32 × 1024 digital output stream.

All measurements were performed at a supply voltage of 0.3 V. Total measured power consumption was 6.6 μW. A 0.3 V full-swing (V_{pp} = 0.3 V) 63-KHz sine wave input signal was sampled at a sampling rate of 3.03 MS/s. Fast Fourier

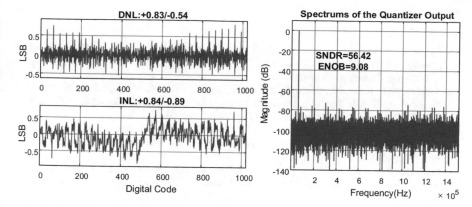

Fig. 6.18 Measured DNL, INL, SNDR, and ENOB

Table 6.1 SAR ADC performance comparison

	ISSCC 2014 [29]	JSSC 2015 [30]	TCAS-II 2016 [31]	JSSC 2017 [32]	JSSC 2017 [33]	ESSCIRC 2017 [34]	This work
DAC topology	CBW	CBW	BWA	CBW	BWA	BWA	BWA
CMOS technology (nm)	65	65	65	65	65	65	90
Unit cap. (fF)	0.55	0.31	26.9	2	2.2	0.14	44
Resolution (bit)	14	13	14	11	12	10	10
Area (mm²)	0.18	0.054	0.28	0.03	0.105	0.048	0.08
Supply voltage (V)	0.8	1.2	0.8	0.7	0.8	1	0.3
Sample frequency (MS/s)	0.032	50	0.01	0.1	0.04	1	3
ENOB (bit)	11.29	11.5	12.5	10.38	10.36	9.17	9.08
Power (μW)	0.352	1000	1.98	0.6	0.375	3.2	6.6
FOM (fJ/conv.-step)	4.4	6.9	34.2	4.5	7.1	5.6	4.065

BWA binary weighted with attenuation capacitor, *CBW* conventional binary weighted, $FOM = \frac{Power}{2^{ENOB} \cdot f_s}$; f_s sample frequency

transform spectrum analysis was conducted and 32,768 data points were collected from the analysis. Figure 6.18 indicates that the measured DNL and INL are +0.83/−0.54 and + 0.84/−0.89, respectively. The SNDR is approximately 56.42 dB and the ENOB is 9.08. The energy efficiency of the power-enhanced SAR ADC is determined to be 4.065 fJ/conv.-step by using the Figure of merit defined in [35].

Table 6.1 summarizes the performance of the proposed SAR ADC and presents a comparison between the performance of the proposed ADC and those of the ADCs presented in previous studies.

References

1. Boulemnakher, M., Andre, E., Roux, J., & Paillardet, F. (2008). A 1.2 V 4.5 mW 10b 100MS/s pipeline ADC in a 65nm CMOS. In *IEEE International Solid-State Circuits Conference-Digest of Technical Papers*, pp. 250–611.
2. Yoshioka, M., Kudo, M., Mori, T., & Tsukamoto, S. (2007). A 0.8 V 10b 80MS/s 6.5 mW pipelined ADC with regulated overdrive voltage biasing. In *IEEE International Solid-State Circuits Conference. Digest of Technical Papers*, pp. 452–614.
3. Brooks, L., & Lee, H.-S. (2009). A 12b 50MS/s fully differential zero-crossing-based ADC without CMFB. In *IEEE International Solid-State Circuits Conference-Digest of Technical Papers*, pp. 166–167.
4. Bernstein, K., Bhushan, M., & Rohrer, N. (2001). On the selection of the optimal threshold voltages for deep submicron CMOS technologies. In *IBM Microelectronics, 1st Quarter*, pp. 29–31.
5. Markovic, D., Wang, C. C., Alarcon, L. P., Liu, T. T., & Rabaey, J. M. (2010). Ultralow-power design in near-threshold region. *Proceedings of the IEEE, 98*, 237–252.
6. Liu, C. C., Chang, S. J., Huang, G. Y., & Lin, Y. Z. (2010). A 10-bit 50-MS/s SAR ADC with a monotonic capacitor switching procedure. *IEEE Journal of Solid-State Circuits, 45*, 731–740.
7. Liu, C.-C., Chang, S.-J., Huang, G.-Y., & Lin, Y.-Z. (2009). A 0.92 mW 10-bit 50-MS/s SAR ADC in 0.13-μm CMOS process. In *2009 IEEE Symposium on VLSI Circuits*, pp. 236–237.
8. Van Elzakker, M., Van Tuijl, E., Geraedts, P., Schinkel, D., Klumperink, E., & Nauta, B. (2008). A 1.9-μW 4.4 fJ/conversion-step 10b 1MS/s charge-redistribution ADC. In *IEEE International Solid-State Circuits Conference-Digest of Technical Papers*, pp. 244–610.
9. Ginsburg, B. P., & Chandrakasan, A. P. (2005). An energy-efficient charge recycling approach for a SAR converter with capacitive DAC. In *Proceedings - IEEE International Symposium on Circuits and Systems*, pp. 184–187.
10. Ginsburg, B. P., & Chandrakasan, A. P. (2007). 500-MS/s 5-bit ADC in 65-nm CMOS with split capacitor array DAC. *IEEE Journal of Solid-State Circuits, 42*(4), 739–747.
11. Chang, Y.-K., Wang, C.-S., & Wang, C.-K. (2007). A 8-bit 500-KS/s low power SAR ADC for bio-medical applications. In *IEEE Asian Solid-State Circuits Conference*, pp. 228–231.
12. Zhu, Y., et al. (2010). A 10-bit 100-MS/s reference-free SAR ADC in 90-nm CMOS. *IEEE Journal of Solid-State Circuits, 45*, 1111–1121.
13. Brenna, S., Bonfanti, A., & Lacaita, A. L. (2016). A 70.7-dB SNDR 100-kS/s 14-b SAR ADC with attenuation capacitance calibration in 0.35-μm CMOS. *Analog Integrated Circuits and Signal Processing, 89*, 357–371.
14. Srams, P., Singh, R., Bhat, N., & Ieee, S. M. (2002). An offset compensation technique for latch type sense amplifiers in high speed low. In *IEEE Transactions on Very Large Scale Integration (VLSI) Systems*, vol. 12, pp. 652–657.
15. Abougindia, I. T., Cevik, I., Ay, S. U., & Zghoul, F. N. (2013). A fast two-step coarse-fine calibration (CFC) technique for precision comparator design. In *Proceedings of the IEEE International Conference on Electronics, Circuits, and Systems*, pp. 153–156.
16. Brenna, S., Bonfanti, A., Abba, A., Caponio, F., & Lacaita, A. L. (2014). Analysis and optimization of a SAR ADC with attenuation capacitor. In *37th International Convention on Information and Communication Technology, Electronics and Microelectronics, MIPRO 2014 - Proceedings*, pp. 68–73.
17. Schinkel, D., Mensink, E., Klumperink, E., Tuijl, E. V., & Nauta, B. (2007). A double-tail latch-type voltage sense amplifier with 18ps setup + hold time. In *IEEE International Solid-State Circuits Conference 2007. Digest of Technical Papers*, pp. 314–316.
18. Pinto, A., & dos Santos Ribeiro Fernandes, J. M. (2013). A study on the offset voltage of dynamic comparators. In *Design of Circuits and Integrated Systems 28th Ed*, pp. 347–352.
19. Jeon, H, & Kim, Y.-B. (2010). A low-offset high-speed double-tail dual-rail dynamic latched comparator. In *Proceedings of the 20th symposium on Great lakes symposium on VLSI. ACM*, pp. 45–48.

20. He, J., Zhan, S., Chen, D., & Geiger, R. L. (2009). Analyses of static and dynamic random offset voltages in dynamic comparators. *IEEE Transactions on Circuits and Systems I: Regular Papers, 56*, 911–919.

21. Butzen, P. F. (2007). Leakage *current modeling in Sub-micrometer CMOS complex gates.* Master Thesis at Universidade Federal do Rio Grande do Sul.

22. Liu, J., Jang, T.-K., Lee, Y., Shin, J., Lee, S., Kim, T., et al. (2014). 15.2 A 0.012 mm² 3.1 mW bang-bang digital fractional-N PLL with a power-supply-noise cancellation technique and a walking-one-phase-selection fractional frequency divider. In *2014 IEEE International Solid-State Circuits Conference (ISSCC)*, pp. 268–269.

23. Yeh, C.-W., Hsieh, C.-E., & Liu, S.-I. (2016). 19.5 A 3.2 GHz digital phase-locked loop with background supply-noise cancellation. In *International Solid-State Circuits Conference (ISSCC)*, pp. 332–333.

24. Chen, Y., et al. (2009). Split capacitor DAC mismatch calibration in successive approximation ADC. In *Proceedings of the Custom Integrated Circuits Conference*, pp. 279–282.

25. Yoshioka, M., Ishikawa, K., Takayama, T., & Tsukamoto, S. (2010). A 10-b 50-MS/s 820-µW SAR ADC with on-chip digital calibration. *IEEE Transactions on Biomedical Circuits and Systems, 4*, 410–416.

26. Saberi, M., Lotfi, R., Mafinezhad, K., & Serdijn, W. A. (2011). Analysis of power consumption and linearity in capacitive digital-to-analog converters used in successive approximation ADCs. *IEEE Transactions on Circuits and Systems I: Regular Papers, 58*(8), 1736–1748.

27. Pelgrom, M. J., Duinmaijer, A. C., & Welbers, A. P. (1989). Matching properties of MOS transistors. *IEEE Journal of Solid-State Circuits, 24*(5), 1433–1439.

28. Ohnhäuser, F. (2008). *Theory and realization of high-end analog-to-digital converters (ADC) based on the principle of successive approximation (SAR)*, PhD Thesis at the University of Erlangen-Nürnberg, Erlangen.

29. Harpe, P., Cantatore, E., & Van Roermund, A. (2014). An oversampled 12/14b SAR ADC with noise reduction and linearity enhancements achieving up to 79.1dB SNDR. *Digest of Technical Papers - IEEE International Solid-State Circuits Conference, 57*, 194–195.

30. Yoshioka, K., Shikata, A., Sekimoto, R., Kuroda, T., & Ishikuro, H. (2014). An 8 bit 0.3–0.8 V 0.2–40 MS/s 2-bit/step SAR ADC with successively activated threshold configuring comparators in 40-nm CMOS. *IEEE Transactions on Very Large Scale Integration (VLSI) Systems, 23* (2), 356–368.

31. Zhang, D., & Alvandpour, A. (2016). A 12.5-ENOB 10-kS/s redundant SAR ADC in 65-nm CMOS. *IEEE Transactions on Circuits and Systems II: Express Briefs, 63*, 244–248.

32. Chen, L., Tang, X., Sanyal, A., Yoon, Y., Cong, J., & Sun, N. (2017). A 0.7-V 0.6-µW 100-kS/s low-power SAR ADC with statistical estimation-based noise reduction. *IEEE Journal of Solid-State Circuits, 52*, 1388–1398.

33. Liu, M., Van Roermund, A. H. M., & Harpe, P. (2017). A 7.1-fJ/conversion-step 88-dB SFDR SAR ADC with energy-free "swap to reset". *IEEE Journal of Solid-State Circuits, 52*, 2979–2990.

34. Bindra, H. S., Annema, A.-J., Louwsma, S. M., van Tuijl, E. J. M., & Nauta, B. (2017). An energy reduced sampling technique applied to a 10b 1MS/s SAR ADC. In *ESSCIRC 2017-43rd IEEE European Solid State Circuits Conference*, pp. 235–238.

35. Craninckx, J., & Van der Plas, G. (2007). A 65fJ/conversion-step 0-to-50MS/s 0-to-0.7 mW 9b charge-sharing SAR ADC in 90nm digital CMOS. In *International Solid-State Circuits Conference. Digest of Technical Papers*, pp. 246–600.

Chapter 7
Summary

This book combines expertise in different fields, including biotechnology and electrical and electronic engineering. The biological signals are usually dedicated to a specific frequency range or a specific voltage range. As engineers, we need to think about how to capture disturbed or noisy biomedical signals and convert them into meaningful biosignals. This intermediate process includes signal amplification (amplifier design), noise removal (filter design), analog-to-digital conversion (analog-to-digital converter design), digital signal processing (algorithm design), and data transmission through human–machine interface (interface circuit design) or wirelessly (frequency synthesizer design, radio frequency circuit design). All these topics belong to electronic engineering. Information brought by biological signals can be used for many applications. For example, rehabilitation aids or equipment can be manufactured through a human–machine interface to improve the quality of life. The implanted devices also allow the patient to immediately apply appropriate feedback, such as cochlear implant to stimulate the auditory nerve or pacemaker to stimulate the heartbeat. These signals can also be wirelessly transmitted to a medical center or emergency center to notify relevant units of immediate responses and preparation so that patients can receive treatment and immediate care.

In Chap. 2, we introduce low-power and low-voltage VLSI circuit design for biomedical applications. With the development of process technology, it brings many benefits, such as smaller area, faster speed, more functions, and lower power consumption. However, it also brings some problems, such as various leakage currents and design difficulties. There are problems we need to face in the design of low-voltage analog circuits, so we have introduced G_M/I_D design methods to overcome them. Some considerations for the design and implementation of nanometer analog circuits are also discussed. How to design digital and analog circuits together with good performance under low power consumption and low voltage is the focus of this chapter.

© The Author(s), under exclusive license to Springer Nature Switzerland AG 2022   215
C.-C. Hung, S.-H. Wang, *Ultra-Low-Voltage Frequency Synthesizer*
and Successive-Approximation Analog-to-Digital Converter for Biomedical
Applications, Analog Circuits and Signal Processing,
https://doi.org/10.1007/978-3-030-88845-9_7

Chapter 3 mainly introduces the integer-N frequency synthesizer by beginning with basic working principles and linear models of frequency synthesizers. It also discusses trade-offs when designing a frequency synthesizer. The working principle of each component of the frequency synthesizer is also presented. Then, we introduce the other types of non-Integer-N frequency synthesizers and include problems and solutions in the realization of the all-digital frequency synthesizer. Finally, a design example of the frequency synthesizer is provided to conclude the chapter.

In Chap. 4, we illustrate a design example of a low-power low-voltage integer-N frequency synthesizer for biomedical applications. The design has been published in *IEEE TBioCAS*. The techniques introduced include curvature phase frequency detector technology to achieve faster lock and lower jitter. The charge pump adopts a two-step switching technique, which greatly reduces the power consumption of the current mirror and reduces noise. Leakage analysis and subthreshold leakage reduction techniques can reduce the reference spurs and jitter of the voltage-controlled oscillator. The dithering technique uses a random method to decompose the spurs of the reference frequency and reduce the average level. Finally, the measurement results are attached to prove its effectiveness.

Chapter 5 introduces the working principle and simulation of analog-to-digital converters. It also briefly explains the working principles of many types of analog-to-digital converters. Next, we discuss the successive approximation analog-digital converters in more details, including algorithms, calibration technology, and the latest development trends. Finally, attentions are paid to the common problems of ADC simulation and testing.

In Chap. 6, we show an implementation example of a low-power and low-voltage analog-to-digital converter for biomedical application. The design has been published in *IEEE TBioCAS*. This research proposes several techniques to improve ADC performance. The proposed pipeline comparator can maintain the advantages of dynamic comparators and reduce the disadvantages of kickback noise. The weighted offset calibration technique is used to calibrate the offset voltage of the comparator without the sacrifice of the operating speed of the traditional calibration circuits. By adding a unity gain buffer, the leakage problem of the bootstrap switch during the hold period is improved. Moreover, the influence of top-plate parasitic capacitance on the capacitance ratio in the digital-to-analog converter is reduced. Finally, the measurement results are attached to prove its effectiveness.

We hope this book can help engineers with electronic and electrical engineering backgrounds to quickly master the knowledge and skills of commonly used analog-to-digital converters and frequency synthesizers. Although the ultra-low voltage brings the benefits of low power consumption, it also introduces design problems and challenges. These challenges will be overcome by electrical engineers with innovative thinking.

Appendix

Matlab Code for Design Example of PLL

```
%PLL design example
N=64;
R1=2.43*10^2;
C1=1.58*10^-8;
C2=3.29*10^-9;
Kvco=200*10^6;
Ip=1*10^-3;

num_lpf=[R1*C1 1];
den_lpf=[R1*C1*C2 C1+C2 0];

num_pd=[Ip];
den_pd=[2*pi];

num_div=[1];
den_div=[N];

num_vco=[2*pi*Kvco];
den_vco=[1 0];

[num1 den1]=series(num_pd,den_pd,num_lpf,den_lpf);
[num_fw den_fw]=series(num1,den1,num_vco,den_vco);
[num_op den_op]=series(num_fw,den_fw,num_div,den_div);
[num_cl den_cl]=feedback(num_fw,den_fw,num_div,den_div);

%
w=logspace(-1,10,200);
%
[mag,phase,w]=bode(num_op,den_op,w);
subplot(311),margin(mag, phase, w);
```

```
%title('Open Loop Frequency Response');
 [mag,phase,w]=bode(num_cl,den_cl,w);
 subplot(312),margin(mag, phase, w);
 title('Closed Loop Frequency Response');
 t=[0:0.0000005:0.00004];
 subplot(313),step(num_cl*4*10^6,den_cl,t);
 title('Frequency Hoppong Response');
 ST=20*10^-6;
 set_time=stepinfo([num_cl*4*10^6 den_cl],'SettlingTimeThreshold',
ST)
```

Matlab Code to Calculate INL/DNL of Sine Histogram Method

```
clear;
close all;

VDD=1; %set supply voltage
A=load('test_result_10.txt'); %The code is based on 10-bit ADC
measurement results
B9=A(:,2); %The measured bit sequence is from MSB to LSB and starts with a
time stamp.
B8=A(:,3);
B7=A(:,4);
B6=A(:,5);
B5=A(:,6);
B4=A(:,7);
B3=A(:,8);
B2=A(:,9);
B1=A(:,10);
B0=A(:,11);

B9=round(B9/VDD); % Normalized with VDD
B8=round(B8/VDD);
B7=round(B7/VDD);
B6=round(B6/VDD);
B5=round(B5/VDD);
B4=round(B4/VDD);
B3=round(B3/VDD);
B2=round(B2/VDD);
B1=round(B1/VDD);
B0=round(B0/VDD);
Daq=B9+2*B8+4*B7+8*B6+16*B5+32*B4+64*B3+128*B2+256*B1+512*B0;
N=size(Daq,1);
Bit=10;
if N==1
  Daq=Daq';
end
% N=size(Daq,1);
MaxCode=2^Bit-1;
```

```
Code=0:MaxCode;
Hdaq=hist(Daq,Code);
Miscode=find(Hdaq<1);
nCode0=Hdaq(1);
%nCode0=max(Hdaq(1:2^Bit/2));
Hdaq(1)=0;
nCodeMax=Hdaq(end);
%nCodeMax=max(Hdaq(2^Bit/2:end));
Hdaq(end)=0;
figure
stem(Code,Hdaq,'Marker','none');
title('HDaq');
% ===============================
C1=cos(pi*nCodeMax/N);    %The largest code, index plus one
C2=cos(pi*nCode0/N);         %The minimum code is 0, index plus one
Offset=(C2-C1)/(C2+C1)*(2^(Bit-1)-1); %the unit is LSB ,it mean
different between idea and real.
Peak=(2^(Bit-1)-1-Offset)/C1;   %the unit is LSB if p2p is 1,then Peak =
0.5
Hsin=(N/pi)*(asin((Code(2:end-1)+1-2^(Bit-1)-Offset)/Peak)-asin
((Code(2:end-1)-2^(Bit-1)-Offset)/Peak));
Hsin=[0 Hsin 0];
hold on;
plot(Code,Hsin);
xlim([0 MaxCode]);
hold off;
% ===============================
Code=0:(size(Hsin,2)-1);
CodeWidth=Hdaq./Hsin;
DNL=CodeWidth-1;
DNL(1)=0;
DNL(end)=0;
max_dnl=max(DNL);
min_dnl=min(DNL);
subplot(2,1,1)
plot(Code,DNL);
title('DNL');
xlabel('Digital Code');
ylabel('LSB');
xlim([0 MaxCode]);
ylim([min_dnl-0.1 max_dnl+0.1]);
text(0.8*(MaxCode),max_dnl,sprintf('max=%4.4fLSB\n min=%4.4fLSB',
max_dnl,min_dnl),'Hor','right')
grid on;
INL(1)=0;
for tmpi=2:(size(DNL,2))
  INL(tmpi)=INL(tmpi-1)+DNL(tmpi);
end

max_inl=max(INL);
min_inl=min(INL);
subplot(2,1,2);
plot(Code,INL);
```

```
title('INL');
xlabel('Digital Code');
ylabel('LSB');
xlim([0 MaxCode]);
ylim([min_inl-0.1 max_inl+0.1]);
text(0.8*(MaxCode),max_inl,sprintf('max=%4.4fLSB\n min=%4.4fLSB',
max_inl,min_inl),'Hor','right')
grid on;
```

Index

© The Author(s), under exclusive license to Springer Nature Switzerland AG 2022
C.-C. Hung, S.-H. Wang, *Ultra-Low-Voltage Frequency Synthesizer
and Successive-Approximation Analog-to-Digital Converter for Biomedical
Applications*, Analog Circuits and Signal Processing,
https://doi.org/10.1007/978-3-030-88845-9

Printed in the United States
by Baker & Taylor Publisher Services